Mathematical Apocrypha

Stories and Anecdotes of Mathematicians and the Mathematical

Library of Congress Catalog Card Number 2002107968

ISBN 0-88385-539-9

Printed in the United States of America

Current Printing (last digit):
10 9 8 7 6 5 4 3 2 1

Mathematical Apocrypha

Stories and Anecdotes of Mathematicians and the Mathematical

Steven G. Krantz
Washington University in St. Louis

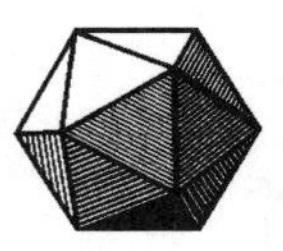

Published and distributed by
The Mathematical Association of America

In memory of
Halsey Royden (1928–1993),
an inspiring storyteller.

The Spectrum Series of the Mathematical Association of America was so named to reflect its purpose: to publish a broad range of books including biographies, accessible expositions of old or new mathematical ideas, reprints and revisions of excellent out-of-print books, popular works, and other monographs of high interest that will appeal to a broad range of readers, including students and teachers of mathematics, mathematical amateurs, and researchers.

777 Mathematical Conversation Starters, by John dePillis

All the Math That's Fit to Print, by Keith Devlin

Circles: A Mathematical View, by Dan Pedoe

Complex Numbers and Geometry, by Liang-shin Hahn

Cryptology, by Albrecht Beutelspacher

Five Hundred Mathematical Challenges, Edward J. Barbeau, Murray S. Klamkin, and William O. J. Moser

From Zero to Infinity, by Constance Reid

The Golden Section, by Hans Walser. Translated from the original German by Peter Hilton, with the assistance of Jean Pedersen.

I Want to Be a Mathematician, by Paul R. Halmos

Journey into Geometries, by Marta Sved

JULIA: a life in mathematics, by Constance Reid

The Lighter Side of Mathematics: Proceedings of the Eugène Strens Memorial Conference on Recreational Mathematics & Its History, edited by Richard K. Guy and Robert E. Woodrow

Lure of the Integers, by Joe Roberts

Magic Tricks, Card Shuffling, and Dynamic Computer Memories: The Mathematics of the Perfect Shuffle, by S. Brent Morris

The Math Chat Book, by Frank Morgan

Mathematical Apocrypha: Stories and Anecdotes of Mathematicians and the Mathematical, by Steven G. Krantz

Mathematical Carnival, by Martin Gardner

Mathematical Circus, by Martin Gardner

Mathematical Cranks, by Underwood Dudley

Mathematical Fallacies, Flaws, and Flimflam, by Edward J. Barbeau

Mathematical Magic Show, by Martin Gardner

Mathematical Reminiscences, by Howard Eves

Mathematics: Queen and Servant of Science, by E.T. Bell

Memorabilia Mathematica, by Robert Edouard Moritz

New Mathematical Diversions, by Martin Gardner

Non-Euclidean Geometry, by H. S. M. Coxeter

Numerical Methods That Work, by Forman Acton

Numerology or What Pythagoras Wrought, by Underwood Dudley

Out of the Mouths of Mathematicians, by Rosemary Schmalz

Penrose Tiles to Trapdoor Ciphers ... and the Return of Dr. Matrix, by Martin Gardner

Polyominoes, by George Martin

Power Play, by Edward J. Barbeau

The Random Walks of George Pólya, by Gerald L. Alexanderson

The Search for E.T. Bell, also known as John Taine, by Constance Reid

Shaping Space, edited by Marjorie Senechal and George Fleck

Student Research Projects in Calculus, by Marcus Cohen, Arthur Knoebel, Edward D. Gaughan, Douglas S. Kurtz, and David Pengelley

Symmetry, by Hans Walser. Translated from the original German by Peter Hilton, with the assistance of Jean Pedersen.

The Trisectors, by Underwood Dudley

Twenty Years Before the Blackboard, by Michael Stueben with Diane Sandford

The Words of Mathematics, by Steven Schwartzman

MAA Service Center
P.O. Box 91112
Washington, DC 20090-1112
800-331-1622 FAX 301-206-9789
www.maa.org

Preface

This book came about in the following unusual manner. Fourteen years ago I read, with considerable pleasure, the book *Who Got Einstein's Office*—an unauthorized history of the Institute for Advanced Study—by Ed Regis. After all, I was a graduate student in Princeton and have been a visiting scholar at the Institute; reading this book was a bit like going home.

I subsequently wrote Regis a letter, pointing out that many of his anecdotes about the denizens of Princeton and the Institute were incorrect. I provided him with the (as best I knew) correct versions of the stories. Regis replied in a very friendly letter, observing that he had been uncertain about many of the stories. But nobody in Princeton would talk to him, so his sources of information were tenuous. He thanked me for the corrections, and suggested that I write a book or an article about mathematical anecdotes.

That I did. In fact my article "Mathematical Anecdotes" appeared in the *Mathematical Intelligencer* in 1990. Response to that article has been almost uniformly enthusiastic. A number of people have sent me letters telling of their personal experiences with von Neumann and Wiener and Lefschetz and many of the other subjects of my anecdotes. I have been sent doggerel and quotations and speculations. I have even been given copies of certain FBI files obtained through the Freedom of Information Act. Over time, I have finally collected enough material so that I could begin thinking about a book. Memoirs of Halmos, Ulam, Rota, and others also provided some grist for my mill. And, anyway, I had some desire to record all these musings in one place, so that they would not be lost. It is in this way that the present volume evolved.

I must, as usual, begin with a standard *caveat*. There is no sense in this work, nor any intention, of making fun of anyone nor of evincing any disrespect. I literally revere the mathematicians who are described in the stories herein. I have learned immensely from all of them; I wish that I were more like them. I revel in knowing tidbits of their lives, and I wish that I knew more. The telling of stories about our heroes is an old tradition that goes back to Homer and even earlier. It is a noble activity to cultivate and to continue.

On reading my manuscript for this book, Constance Reid said

> I have always thought of mathematical anecdotes as pretty much unique to mathematicians. The best ones, for me, are those that encapsulate a mathematician's character or personality with all the economy of a formula.

I have spent my entire adult life hanging around academics, and have certainly never encountered a group that is so hell-bent on telling stories about each other as are mathematicians. With this book I plant my flag as a storyteller.

The scholarly reader will note that the dictionary (*Webster's Third New International Dictionary*, unabridged) meaning of the word "apocrypha" is

> Quasiscriptural noncanonical or deuterocanonical books of doubtful authorship and authority... Writings or statements of doubtful or spurious authorship.

What mathematician can resist a word whose definition uses the revered term "canonical?" My title is *not* meant to suggest that the stories contained herein are in any sense unreliable or fabricated. Rather, the title is meant in part to attract the potential reader's attention and in part to suggest an entrée to some fun. Most of the stories here are in fact verifiable, and have been checked (in the fashion of investigative reporters) with other witnesses.

The theme of this book is strictly mathematical. Some of the stories, however, are about people who adhere to mathematics but cannot strictly be called mathematicians. I include here Bertrand Russell, Alfred North Whitehead, Albert Einstein, and others. I think that the spirit of these ancillary stories is still true to the purpose of this collection: to immerse the reader in mathematical culture, to develop an appreciation for mathematics and mathematicians, and to inform and amuse.

In selecting the pieces contained herein, I have been guided by a few principles. First, I am only interested in stories about *people*. Simply recording quotations or witty aphorisms will be left for another venue. I also only want to include pieces that have intrinsic interest, and that will mean something to a broad mathematical audience (not just to those who are named in the story). I want these to be stories that the reader will enjoy and then want to repeat to someone else. That is the spirit of the anecdote.

The stories here depict mathematicians who are joyous, mathematicians who are serious, mathematicians who are triumphant, mathematicians who are foolish, mathematicians who are human. They are a celebration of the mathematical life and the people who live it. I have strenuously avoided the telling of stories that are mean-spirited or critical or that depict people in a bad light. I want these stories to make people happy, not sour.

The telling of stories acquaints beginning mathematicians with our culture and with our values. It shows them how we think and how we operate. These stories are not always flattering; in a few cases they are not verifiably true. All I can say about some stories in this collection is that somebody told them to someone else at one time or another. Many of them were told directly to me, or were events that I witnessed. Some others I got from very reliable witnesses. It would serve no good purpose to sort out which are which. I find them all great fun; they constitute a celebration of the life of mathematics. I hope that the reader will enjoy them as much as I have.

I am happy to thank Lynn S. Apfel, Edwin Beschler, Brian Blank, Keith Dennis, John Ewing, Matelda Colautti Fichera, Ron Freiwald, Mike Jury, John McCarthy, and many other friends for reading drafts of this manuscript and for offering emendations and corrections. Ed Dunne contributed some of the stories from his own first-hand experience. Jerry Alexanderson read the entire manuscript—many times!—with special care and offered many suggestions and corrections. Constance Reid also gave freely of her time and expertise, offering many useful suggestions and criticisms; she dipped into her well of historical knowledge and gave me valuable information—and more stories. Jerry and Connie have my heartfelt thanks. A great many resourceful people helped me to find dates for the characters mentioned herein; these include Jerry Alexanderson, David Azzolina, Ed Dunne, Julie Honig, Barbara Luszczynska, and Randi Ruden. Randi Ruden contributed sensitive guidance as to what to include and what to omit. The MAA publications committee provided a variety of helpful remarks and suggestions that certainly sharpened my focus on this project. Don Albers was the perfect editor and publisher, working quietly and consistently to make every step of the publishing process both comfortable and effective. Both Don Albers and Jerry Alexanderson did yeoman duty in identifying and collecting photographs for the book. The photos help to bring the characters to life, and add to the piquancy of many of the stories.

All remaining errors and inaccuracies are of course the responsibility of this scribe. I will be happy to learn of all corrections and criticisms, in the hope of publishing a more accurate future edition.

SGK
St. Louis, MO

Table of Contents

Great Foolishness

George Mackey (1916–) was with a group of other mathematicians on the free afternoon of a conference in New Mexico. They were engaged in a hike in the desert. The air was crisp and dry, the sky was blue and without clouds, the temperature was near 100°, the hike was brisk.

Mackey was wearing rain galoshes. Walter Rudin (1921–) asked him why, and Mackey replied, "Well, this is one less thing to worry about."

Presumably it was his preoccupation with mathematics that caused Stefan Bergman (1895–1977) to appear to be out of touch with reality at times. For example, one day he went to the beach in northern California with a group of people, including a friend of mine who told me this yarn. Northern California beaches are cold, so when Bergman came out of the water he decided that he had better change into his street clothes. As he wandered off into the parking lot, seeking the car where he could get his clothes and change, his friends noticed that he was headed in the wrong direction. But they were used to this sort of behavior and paid him no mind. In a while, Bergman returned—clothed—but plainly not in his own clothes. He exclaimed, "You know, there is the most unfriendly woman in our car!"

Logician Alfred Tarski (1902–1983) was actually named Teitelbaum. He changed his name to protect himself from anti-semitic prejudices. He hated to be called Teitelbaum. In 1958 Abram Besicovitch (1891–1970) ran into him and immediately began calling him "Teitelbaum." Tarski got angry and

Alfred Tarski

demanded that he not do so. He said, "You called me Teitelbaum the last time you saw me, whenever that was." Besicovitch said, "Yes, it was at the International Congress in 1954."

Abram Besicovitch was of the old school, and he lived in times that are hard for us today to understand. In particular, he taught in England at a time when long distance phone calls were considered to be a real luxury. It was quite unusual for people to call ahead when going to visit someone. They would just jump in the car and go and hope to find the person at home.

And this is what Besicovitch did. He drove for a couple of hours, and was overjoyed to find his old friend at home. They embraced warmly, and were soon involved in a detailed discussion of mathematics. After a time, the friend said, "Well, Abram. It is lunch time and you must be hungry. Let us eat." Which they did. After lunch they resumed their talk about mathematics. Five or six hours later the friend said, "Well, Abram it is time for dinner. Won't you join me?" Besicovitch readily assented. "But," the friend said, "hadn't you better phone your wife? She is probably worried about you. Perhaps she is already preparing dinner at your home." Besicovitch said, "No, she is not worried. She is waiting in the car."

Pete Casazza (1945–) is an impish mathematician at the University of Missouri in Columbia. One semester he was assigned to teach a large calculus lecture. This was a task that Casazza had assumed many times before, and he was pretty good at it, but he was also tired of it. So he decided to take a new approach. He arranged for a "ringer"—someone who was not Casazza but who would pretend to be Casazza—to meet the class on the first day. Casazza sat in the audience near the front. The lecture began, and Casazza, affecting to be a student, peppered the lecturer with questions and comments. He found many faults with the presentation. The lecturer became increasingly frustrated and irritated. Finally, in exasperation, the lecturer threw down his chalk and cried, "All right. If you think you can do a better job then *you* teach the class." He then stormed out of the room. So Casazza took over.

G. H. Hardy (1877–1947) and J. E. Littlewood (1885–1977) discussed the concept of stage fright. They agreed that, for a lecture in front of the Royal Society, or a lecture at a foreign university, stage fright was not a problem. You knew what you were talking about, you were a ranking expert, you were among equals, and you could get up and strut your stuff. But in front of a calculus class, first lecture of the Fall term, there was definitely stage fright.

One day Shizuo Kakutani (1911–) was teaching a class at Yale. He wrote down a lemma on the blackboard and announced that the proof was obvious. One student timidly raised his hand and said that it wasn't obvious to him. Could Kakutani explain? After several moments' thought, Kakutani realized that he could not himself prove the lemma. He apologized, and said that he would report back at their next class meeting.

After class, Kakutani went straight to his office. He labored for quite a time and found that he could not prove the pesky lemma. He skipped lunch and went to the library to track down the lemma. After much work, he finally found the original paper. The lemma was stated clearly and succinctly. For the proof, the author had written, "Exercise for the reader." The author of this 1941 paper was Kakutani.

André Weil

Norbert Wiener's (1894–1964) father was a distinguished linguist. Wiener followed in his father's footsteps, to the extent of learning many languages. He was particularly proud of his ability with Chinese.

Wiener was once invited to lecture in China. He wanted to begin his first lecture with a little Chinese, so he spoke some words. The Chinese listened very politely. It was later observed that what Wiener actually said was "The cow is green."

One day Norbert Wiener, the harmonic analyst, André Weil (1906–1998), the algebraic geometer, S. S. Chern (1911–), the Chinese geometer, and some others were riding in an elevator at MIT. Weil also knew some Chinese, and he knew that Wiener did too. So Wiener and Weil jabbered away in Chinese during the rather long elevator ride. After they got off, Chern turned to a graduate student and said, "Can you please tell me what language they were speaking?"

Wiener was invited to a dinner party in the Boston area. He was by far the most distinguished guest. Therefore the hostess took great pains to find out what dishes Wiener liked and planned the meal around them. When the guests sat down to dinner, Wiener took out a bag of peanuts and announced that *that* was what he was going to eat.

As we have noted elsewhere, Wiener was quite a celebrity around MIT. Students were in awe of him. Therefore, when one of his students spied Wiener in the post office, the student wanted to introduce himself to the famous professor. After all, how many MIT students could say that they had actually shaken the hand of Norbert Wiener? However, the student wasn't sure how to approach the famous savant. The problem was aggravated by the fact that Wiener was pacing back and forth, deeply lost in thought. Were the student to interrupt Wiener, who knows what profound idea might be lost? Still, the student screwed up his courage and approached the great man. "Good morning, Professor Wiener," he said. The professor looked up, struck his forehead, and cried, "Wiener! That's the word."

In 1984, I (Steven G. Krantz, 1951–) visited the University of Oslo in Norway to give a colloquium talk. I was told at the time that the country was in social upheaval—for two reasons that could be laid at the feet of the Americans. One is that the custom in Norway had been for everyone to do their shopping on Saturday mornings. People enjoyed the friendly hustle and bustle, liked meeting their neighbors, and took pleasure in buying groceries and other necessities. But now the tradition was broken because everyone was staying home to watch reruns of the television show *Dynasty*. The other catastrophe was that the custom in Norway had been for everyone to go to church on Sunday morning. But now—instead—everyone was lining up at the new MacDonald's in downtown Oslo.

Yuri V. Uspenski told about the visit of an education commissar to the university where he was teaching in Russia after the Revolution. This commissar asked him about a course the math department taught called "the theory of ideals." Uspenski hurriedly informed him that the course had already been changed. The new course was called "the theory of classes."

There is a famous mathematician, son of one of the really eminent mathematicians of this century, who teaches at a large public university in this country. For reasons of discretion, we shall call him "Professor *X*." One day this scholar was teaching calculus. A student raised his hand and queried, "Professor *X*, what is 'infinity'?" The professor nodded seriously and said, "It is like a long line that never stops," and he proceeded to apply his chalk to the blackboard, walking in a determined manner toward the side of the room. When he reached the window he kept going, through and out the window. Suddenly there was no teacher in the classroom. The students sat for several moments in bewildered silence, until finally one of their number went over to the window to determine what had become of Professor *X*. In fact the student could see him two floors below, spread-eagled in the bushes (and unharmed).

One day Warren Ambrose (1914–1995) of MIT came to class with one shoelace tied and one untied. His students asked him whether he knew that his right shoelace was untied. Without hesitation, Ambrose adopted a quizzical look and said, "Oh, my God. I tied the left one and thought the other must be tied by considerations of symmetry."

After John Nash (1928–) won the Nobel Prize for Economics in 1994, a small ceremony was held in the Fine Hall Common Room at Princeton University. Nash was prevailed upon to make a few remarks. His first was, "I hope that getting the Nobel will improve my credit rating, because I really want a credit card."

A bright young mathematician—who shall remain nameless for reasons of discretion—had his first job at the Mittag-Leffler Institute and his second at Yale. Both are fairly intimate venues for doing mathematics. People must co-exist cheek-by-jowl, and extra care must be taken to be considerate of other people in the building. Well, our friend was young and brilliant and extremely boisterous. He had the nasty habit of jumping up in the middle of a seminar, running to the front of the room, grabbing the chalk from the lecturer, and taking over the proceedings. Whether he was right or not (and he frequently *was* right), people found that this behavior grated on their nerves.

A more senior member of the audience (both at Mittag-Leffler and at Yale)—a cynical Frenchman—developed the habit of drawling in disgust, "Take him to the prostitutes!"

Marco Abate (1962–) is a talented complex analyst who works at the University of Rome. Marco is an energetic and active guy, with many papers and books to his credit. One of his lifelong dreams, which he has finally realized, has been to publish a comic book. Now the comic book is out—published by a major commercial publisher and illustrated by a professional graphic artist. The super-hero in the comic book is "Professor Krantz" (yes, *that* Professor Krantz) and he fights the evils of fractals (yes, *those* fractals). Let Benoit Mandelbrot (1924–) beware.

Richard Bellman (1920–1984) was quite a distinguished mathematician in his day, the founder of *The Journal of Mathematical Analysis and its Applications*. He liked to say that he got tired of responding to the social question, "And what do you do for a living?" by saying he was a mathematician. This usually got a tiresome response, or no response at all. So he

Richard Bellman

often told people that he was a tennis coach. At parties, this proved to be a much more salubrious social touch.

Princeton University has a bad case of celebrity-itis, as do many prestigious institutions. After all, the rich and famous are a source of additional fame and also of substantial donations. When I was a graduate student there, the daughter of Ferdinand and Imelda Marcos was a student. She had special permission to be accompanied by arms-bearing bodyguards to all her classes.

Years later, Brooke Shields went to Princeton. Rumor had it that Princeton was the only top school that was willing to admit her without examining her *bona fide*s. A friend of mine—a math major—really had the hots for Brooke, and was dying to meet her.

One day, walking across campus, he spied her talking to one of his professors. This was his big chance. He walked over and joined the conversation, expecting to be introduced. The professor and Brooke Shields nodded to my friend, but he was never inducted into the chitchat and never presented to the famous model. Eventually Brooke Shields walked off and the professor turned in an embarrassed state to my friend. "I'm sorry," he said. "I would have introduced you, but I can never remember her name."

A few years ago a Princeton graduate student in mathematics, Milton Babbitt, won a forty-six year court battle. The math department had refused to recognize his thesis as worthy of a degree, and he had been fighting the decision in various legal venues ever since. In the end, the student won and was granted his Ph.D. The subject of the thesis was some variant of the twelve-note musical scale. The judgment of the student and of the press was that this material was far ahead of its time and too difficult for the Princeton faculty—they just couldn't understand it nor appreciate it.

John Pierpont Morgan (1837–1913) was a student of David Hilbert. Of course this is one and the same "J. P. Morgan" who was to become a world-famous financier. At the conclusion of his studies, teacher Hilbert told him that he'd best return to money matters, where at least he knew what he was doing.

There is a unique book sitting on the shelf of the Math Library at UCLA. Admittedly, it is a *samizdat* (self-published) book. But it is typed and bound and sits with perfect dignity alongside many another august tome. It is

Sex, Crime, and Functional Analysis
Part I: Functional Analysis
by J. D. Stein

Not long ago an algebra book was getting ready to go to press in Turkey. At the last moment, the government intervened and mandated that certain letters not be used as variables in the book, as they happened to be the same letters as were commonly used in the acronym for the revolutionary junta in Turkey. The government suggested alternative letters, which were duly incorporated.

A colleague of mine was traveling to a large state university to give a colloquium talk. Naturally the university was located in a rural area, and the last segment of his trip was on a small plane. It was the dead of winter, and there was ice and sleet everywhere. Flying was treacherous, and the passengers were very nervous. In the middle of the flight, the pilot put the plane on auto-pilot, came back among the passengers, and asked to borrow a credit card. People were quite puzzled, but someone eventually handed over his American Express Card. The pilot said, "Thank you" and took the credit card up into the cockpit. He then reached out the window and used the card to scrape ice off the windshield of the aircraft. This story has two lessons: that this aircraft was not very well equipped, and that this particular airline did not pay its pilots enough so that they could qualify for their own credit cards.

My colleague Nets Katz (1972–) spent a semester at UCLA. There was some political tumult on campus because the University of California system was in the process of eliminating the Affirmative Action program. One day students were picketing the administration building. One student carried a hand-lettered sign that said "Education is a Privilege, Not a Right." Unfortunately, the word "privilege" was misspelled.

Andrew Wiles

After Andrew Wiles (1953–) proved Fermat's Last Theorem he became quite a celebrity. News crews were all over the Princeton Math Department for several days. Wiles received many unusual phone calls—at least unusual for a mathematician. One was from the Gap stores, who wanted him to advertise their jeans. Another was from newscaster Barbara Walters's assistant, who invited Wiles to appear on Walters's television show. Andy's response was, "Who is Barbara Walters?" The assistant informed him, and said she would get back to him with further information. When they phoned back they told Andy that they had decided not to interview him after all; they were going to do Clint Eastwood instead.

An account of the evolution of teaching, in view of the "New Math" and the "New New Math," during the past thirty years has received wide attention. The view is of the transmogrification of a particular mathematical problem with changing social mores. One version of it is as follows:

The 1960's: A peasant sells a bag of potatoes for $10. His costs amount to 4/5 of his selling price. What is his profit?

The 1970's: A farmer sells a bag of potatoes for $10. His costs amount to 4/5 of his selling price, i.e., $8. What is his profit?

The 1970's (new math): A farmer exchanges a set P of potatoes for a set M of money. The cardinality of the set M is equal to 10 and each element of M is worth $1. Draw 10 big dots representing the elements of M.

The set C of production costs is composed of 2 big dots less than the set M. Represent C as a subset of M and give the answer to the question: What is the cardinality of the set of profits? (Draw everything in red.)

The 1980's: A farmer sells a bag of potatoes for $10. His production costs are $8 and his profit is $2. Underline the word "potatoes" and discuss with your classmates.

The 1990's: A kapitalist pigg undjustlee akires $2 on a sak of patatos. Analiz this tekst and sertch for erors in speling, contens, grandmar and ponctuassion, and ekspress your vioos regardeng this metid of getting ritch.

I once had a friend named Uri who hailed from Uruguay. He was a graduate student in Engineering at the State University of New York at Buffalo. He was also a great fan of mathematics. One day Uri decided to take the short drive to see Niagara Falls. Everyone knows that the best view of the Falls is from the Canadian side, so he drove across to enjoy the scenery. When it came time to return, he realized that he had not brought his papers with him. All he had was a New York Driver's License. He decided to fake it and claim that he was a U.S. citizen. It should be noted that Uri's English was superb; he could have been a television newscaster. So he thought he could pull this off. Sure enough, the border guards began asking him various routine questions and he handled them with elan and grace. It was clear that there was going to be no problem here. Finally they said, "Just one more question. Where were you born?" Uri smiled and said, "Booofalo." He was detained quite a bit longer while he explained his way out of that one.

Q: What is the difference between a dog scratching at your door and a full Professor of Mathematics scratching at your door?

A: If you let the dog in, then he stops whining.

Kuniko Weltin

Kuniko Weltin, wife of Berkeley mathematician Hung-Hsi Wu (1940–), sent me the following proof (oriented toward the teenage male) that "Girls are Evil."

First, girls take both time and money. So

$$\text{Girls} = \text{Time} \times \text{Money} \tag{1}$$

But time is money, hence

$$\text{Time} = \text{Money} \tag{2}$$

Substituting (2) into (1) yields

$$\text{Girls} = \text{Money}^2. \tag{3}$$

Finally, money is the root of all evil. We conclude that

$$\text{Money} = \sqrt{\text{Evil}} \tag{4}$$

We substitute (4) into (3) to obtain the desired result.

There are many ways to give a math lecture—say a colloquium talk. The most dry and desultory way to do so is to give a few definitions, state a theorem, and then prove it. Perhaps most mathematicians will carry out the task of giving an invited lecture by following this paradigm. There are more daring, and perhaps more entertaining, ways to proceed. One can instead paint a broad picture, peppered with examples, stories of things tried, little jokes, and salty lessons. Other speakers will present a few profound examples and endeavor to draw pithy conclusions. Yet another way to proceed—and one that can be quite arresting—is to present an N-step program for proving some big conjecture (say the Riemann hypothesis) and then to make witty comments on and prove relations among the various steps. When done well, this method can be both enlightening and entertaining. A friend of mine does not think much of this last approach. In moments of irritation, he likes to say, "I have a twenty-five step program for proving the Riemann hypothesis. You count to twenty-four and then you prove the Riemann hypothesis."

I once had a graduate student from Malaysia who was very bright, but quite eccentric. He was a Hindu, and claimed to have no friends and no hobbies. He dined each evening with the Hare Krishnas just because it was the source of a free vegetarian meal, and he didn't mind chanting nonsense for a couple of hours just to pay the piper. This student worked on mathematics all the time. One day he came to me and said, "I want to learn about harmonic analysis. What should I do?" Without hesitation I said, "Read Katznelson's book, *Introduction to Harmonic Analysis* (Yitzhak Katznelson, 1934–). That will give you a terrific introduction." Two weeks later the student was back. He said, "I am reading the statements of the theorems in Katznelson's book. Do you think I should read the proofs?" I stifled a number of tart rejoinders and said, "Yes, the proofs are important. Please read the proofs." Two weeks later he was back. Now he said, "I am reading the statements of the theorems and the proofs in Katznelson's book. Do you think I should look at the exercises?" Ahem. The exercises in Katznelson's book are one of the world's great treasures. Doing them turned me from a tyro into a mathematician. My answer was instantaneous and unwavering: "Of course. Do some of Katznelson's exercises." Two weeks later the student was back. He said, "I am working on one of the exercises from Katznelson's book. I have a very good attack on this problem. I have it all laid out. There is just one little detail that I cannot nail down. Perhaps you can give me a hint." The exercise was to show that a continuous, multiplicative homomorphism from the circle group $\mathbb{T}$ into the multiplicative complex numbers $\mathbb{C}^* \equiv \mathbb{C} \setminus \{0\}$ must in fact have image lying in the circle. This is, of course, the first step to classifying all the characters of the circle group. It's a great exercise. So I said, "OK. Tell me your attack on the problem and I'll give you a hint." He said, "Suppose not. Then I don't know what to do."

Tommy (Charles Brown) Tompkins (1912–1971) was a professor of mathematics at UCLA. He was known to tipple—during the day—sometimes to excess. Unfortunately one day he was sitting at his office desk and flipped over backwards, hitting his head. He was found dead by the janitor a few days later.

Tompkins had an unusual way of meeting his office hours. He would sit at the desk with his door open. If a student appeared, he would crawl under the desk until the student went away. Worked like a charm.

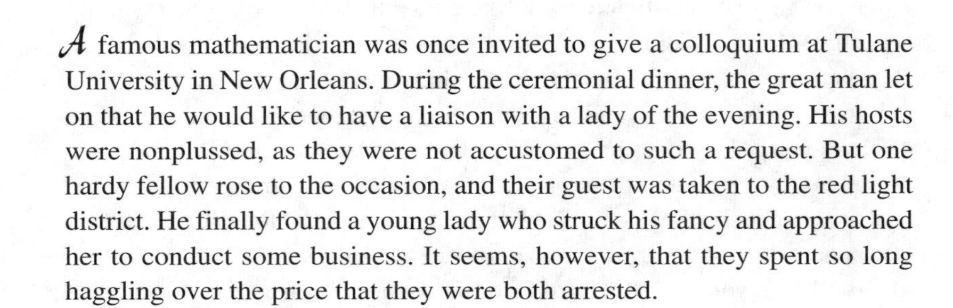

A famous mathematician was once invited to give a colloquium at Tulane University in New Orleans. During the ceremonial dinner, the great man let on that he would like to have a liaison with a lady of the evening. His hosts were nonplussed, as they were not accustomed to such a request. But one hardy fellow rose to the occasion, and their guest was taken to the red light district. He finally found a young lady who struck his fancy and approached her to conduct some business. It seems, however, that they spent so long haggling over the price that they were both arrested.

In the 1980's, the math department at Rice University was in the same building as the Army Reserve Officers Training Corps (ROTC) program. Space was at a real premium at the time, and this strange juxtaposition was one result. One day, while considering the space problem, Professor Bill Veech (1938–) asked, "What is the Army doing in our building? Why aren't they all down in Nicaragua where they belong?"

I have done a good deal of work on pseudoconvex domains in $\mathbb{C}^n$. Thus it was with some pleasure that I finally arranged to get a vanity license plate for my car that says "PSEUDO." The day I got the new plate, I drove to a car wash so that a sparkling car would show off the new adornment. When the final touches were being put on the car, the man polishing the hood asked me what that word was on the license plate. I said, "That's 'pseudo.'" "Well, what does that mean?" queried the car washer. I advised him that it

David Hilbert

means "fake." He immediately turned to the guy next to him who was wiping the windshield and said, "Hey, man. You're *pseudo*."

David Hilbert once attended the funeral of a student of his who died at a tragically young age. He was asked to deliver a few words. Hilbert began by saying what a fine young man the deceased had been. He was very talented and showed a lot of promise. In fact he had been working on a very interesting problem: "Let $\varepsilon > 0$, …" and Hilbert launched into a detailed mathematical disquisition.

Alfred Errera (1886–1960) was a student of Edmund Landau (1877–1938). Errera was quite wealthy, and the dinner parties that he gave were extravagant. It was a great treat to be invited—fancy food, many courses, expensive wines, elegant company, and many servants. One evening Errera gave a dinner party in honor of Paul Lévy (1886–1971), who was notoriously absent-minded. The next day, Lévy and Errera met on the street and the latter, who was unctuously polite, said, "I had great pleasure last evening." Lévy said, "Ah, and where were you last evening?"

J. J. Sylvester (1814–1879) gave a commencement address at Johns Hopkins. He began by remarking that mathematicians were not any good at that sort of thing because the language of mathematics was antithetical to general communication. That is to say, mathematics is very concise: one can express pages of thought in just a few symbols. Thus, since he was accustomed to mathematical expression, his comments would be painfully brief. He finished three hours later.

Dennis Sullivan (1941–) is a great mathematician and a wonderful guy, but he can be quite daring when the occasion presents itself. He was once visiting the math institute IMPA in Brazil. Sullivan's son from an earlier marriage accompanied him, and Dennis had in fact engaged in difficult negotiations with his former wife to get permission to take his son on this trip. In order to visit his new girlfriend's family, Sullivan and his son had to negotiate a jeep trip through fairly desolate, and not very safe, territory. They arrived at a "checkpoint," where armed guerrillas stuck rifles into all the windows and questioned them sharply. At one point the armed men demanded that Sullivan and his son get out of the vehicle for more extensive interrogation. Sullivan reasoned that if they abandoned the car then they were done for. He also determined that if he gunned the engine and they sped away then, since the rifles were already lodged in the windows, the desperadoes would not be able to get a shot off. So he told his son to duck and off they went. Unfortunately, one of the soldiers did shoot and hit Sullivan in the shoulder. Dennis arrived at his destination bleeding, and he took great pains to cover up the damage. Meanwhile, his son was supposed to be phoning his mother every day. Sullivan had to enjoin him not to mention the arms-bearing incident to his mother until he was safely back on U.S. soil.

Because of his many splendid achievements in algebraic topology, including the invention of rational homotopy, Dennis Sullivan landed a permanent position at the Institut des Hautes Etudes Scientifiques (IHES) in Bures-sur-

Yvette, France. At one point he developed an interest in complex dynamics so decided to attend the analysis seminar at the Université de Paris-Sud in neighboring Orsay. In those days, the Grand Séminaire Analyse Harmonique in Orsay was quite a formal affair, run by several older French mathematicians clad in black suits and chain-smoking Gauloises. Sullivan sat in the back, dressed in faded blue jeans and boots. He would have his feet up on the seats, and would lounge in an impudent and idle manner. But he was listening intently.

If the speaker's linguistic abilities were at all up to it, then any presentation in the Grand Séminaire Analyse Harmonique was mandated to be delivered in French. This was never a point of discussion. It was a rule. One day, Sullivan was paying particularly rapt attention to a rather technical talk (in French) on Fourier analysis. In a pronounced Texas drawl he exclaimed, "I can take Fourier analysis, and I can take French, but I can't abide the two together." There was a hurried confab among the black-suited Frenchmen who ran the show. Then, with some sadness in his voice, the leader said to the speaker, "You may speak English."

When I was a graduate student at Princeton, Dennis Sullivan visited from IHES to give a talk. He spoke about the Poincaré conjecture, a very hot topic at the time. Recall that the problem is to show that any closed manifold with the homotopy of the sphere must in fact be homeomorphic to the sphere. In dimension 3, any closed 3-manifold (should such exist) with the homotopy of the sphere which is *not* homeomorphic to the sphere is called a *fake 3-sphere*. Sullivan was going to explain to us how he could construct a moduli space for all fake 3-spheres that was in fact an algebraic variety. This was exciting stuff. And his talk was dazzling. He had commutative diagrams and bundles and arrows all over six blackboards. He was spinning out his tale at blinding speed, and few could begin to follow the details. At one point Bill Browder, getting more and more frustrated, pointed to a particular total space on the upper right and asked, "What does that mean?" Sullivan looked at him in an exasperated fashion and said, "It means 'don't worry.'"

Well, Sullivan ran overtime. *Way* overtime. The professor who was scheduled to teach the next class in that room—someone from another department—stood in the back wringing his hands and making various polite but desperate signals at Sullivan. Sullivan ignored him. Finally the guy said something. Dennis Sullivan aimed his laser gaze at the man and exclaimed (in a thick Texas accent), "I flew all the way from Paris to give this talk and by goddamn I'm going to give it." And he did.

The Preface to *States of Matter*, a recent text on statistical mechanics by David L. Goodstein, reads as follows:

> Ludwig Boltzmann, who spent much of his life studying statistical mechanics, died in 1906, by his own hand. Paul Ehrenfest [Boltzmann's student], carrying on the work, died similarly in 1933. Now it is our turn to study statistical mechanics.

Some graduate students at UCLA once wrote a satirical version of a qualifying exam. Some of the questions that they concocted were these:

1. Let B be a 1956 Buick. Find an imbedding of B into projective 12-space in which B decomposes naturally into the direct sum of two lower-priced models. What is the fundamental group of B?
2. Let F be an almost everywhere, locally upper semi-continuous Boolean idempotent. Show that F is Garfield differentiable. Answer this question in Swedish.
3. Let $\mathcal{H}$ be a Hilbert space. Let $\mathcal{H}$ have prime codimension when viewed as a corneal [sic] variety over itself. Show that $\mathcal{H}$ has Hardly's property, i.e., that $\mathcal{H}$ is Sizemore complete and hardly anything holds in $\mathcal{H}$.
4. In 1872 Frobenius showed that all groups of order 21 are wreathed and thus trivial. The following year Cayley showed that any group whose center is of order 32 is wreathed in the sense of Frobenius. Using only elementary principles (e.g., Hilbert's Satz 47, Jensen's manifold, Kummer's Third) show that Cayley deserves no credit for this discovery.
5. State and prove the Prime Number Theorem, carefully indicating each time that you use commutativity of the reals.
6. Let P be the "big parabola," that is, the set of points on the graph of $f(x) = x^2$ within seven miles of the origin. Build a full scale model of P with materials you have at hand.
7. Using only the symbols #, &, $, @, %, state and prove Gödel's incompleteness theorem. You must use a typewriter.

Kurt Gödel

Logician Kurt Gödel (1906–1978) was an eccentric and unworldly man who frequently needed to be protected from the vicissitudes of life by his friends. After Gödel had lived in this country for many years, he was persuaded to become an American citizen. He therefore began studying for the citizenship exam.

Unfortunately, as soon as Gödel began reading the U. S. Constitution, he discovered troubling logical loopholes. This insight cast him into deep distress. John von Neumann (1903–1957)—Gödel's colleague at the Institute for Advanced Study—was finally called in to convince Gödel that if you looked at things the right way then there would be no logical inconsistency.

Albert Einstein (1879–1955) and the economist Oskar Morgenstern (1902–1976) were the ones who chaperoned Gödel to the hearing for his citizenship application. The judge was overwhelmed by this opportunity to talk to Einstein, and they conversed at length about events in Nazi Germany. Finally, as an afterthought, the judge turned to Gödel and said, "But of course from your reading of the Constitution you now know that nothing like that could happen here." "As a matter of fact," Gödel began—but then, under the table, Morgenstern kicked Gödel. So Gödel got his citizenship after all. [By some accounts, Morgenstern had to do a heck of a lot more than just kick Gödel. Fortunately, he was equal to the task.]

On another occasion, Gödel was required to fill out a bureaucratically and rather cryptically designed draft questionnaire. He studied the form and became increasingly befuddled and frustrated. Instead of giving hasty "yes" and "no" answers, as most of us would do, Gödel wrote a long and involved essay for each question, explaining that if the question meant *A,* then the answer was *X,* while if the question meant *B,* then ....

Gödel was always quite phobic, and something of a hypochondriac. He was sometimes committed to institutions for depression and exhaustion. He avoided human contact—handshakes and other shows of affection. He could be seen, at tea and other social occasions, weaving through the crowd in a strange dance designed to avoid touching other people. Gödel died in 1978; he was convinced that people were trying to poison him, and he starved himself to death.

J. D. Tamarkin

Freeman Dyson (1923–) of the Institute for Advanced Study tells us that, in 1910, the mathematician Oswald Veblen (1880–1960)—a founding member of the Institute—and the physicist James Jeans (1877–1946) were discussing the reform of the mathematics curriculum at Princeton University. Jeans argued that they "may as well cut out group theory," for it "would never be of any use to physics." Of course, today, group theory is central to many parts of physics, such as quantum mechanics. Fortunately, Jeans's advice was not taken.

Once on a Ph.D. oral J. D. Tamarkin (1888–1945), after whom Brown University now has named a research Assistant Professorship, asked the candidate about the convergence properties of certain hypergeometric series. Somewhat churlishly, the student replied, "I don't remember, but I can always look it up if I need it." Tamarkin was displeased. He said, "That doesn't seem to be the case, because you sure need it now."

Paul Halmos (1916–) and John von Neumann had many ongoing jokes, one of which concerned a story of "creative survival." According to von Neumann, a little Jewish farm boy named Moyshe Wasserpiss emigrated to Vienna and became a successful businessman. He changed his name to Herr Wasserman. Going on to Berlin and to even greater success and fortune, he became Herr Wasserstrahl. Reaching another social level a few years later, he graduated to Viscount von Wasserstrahl. Finally in Paris and still more prosperous, he is now Baron de la Fontaine.

Albert Einstein was Gödel's closest personal friend in Princeton. For several years, Einstein, Gödel, and Einstein's assistant Ernst Straus (1922–1983) who later moved to UCLA and specialized in combinatorial theory, would lunch together. During lunch they discussed diverse non-mathematical topics—frequently politics. One notable discussion took place the day after Douglas MacArthur was given a ticker-tape parade down Madison Avenue upon his return from Korea. Gödel came to lunch in an agitated state, insisting that the man in the picture on the front page of the *New York Times* was not MacArthur but an impostor. The proof? Gödel had an earlier photo of MacArthur and a ruler. He compared the ratio of the length of the nose to the distance of the tip of the nose to the point of the chin in each picture. These were different: Q.E.D.

Straus told the story of the day that he and Einstein finished work on a paper. They looked for a paper clip to bind it together. After shuffling through several drawers, they finally found one lone clip. But it was so bent and mangled that it could not be used. So then they began looking for a tool to straighten the forlorn paper clip. Scrounging through more drawers, they finally found a full box of brand new paper clips. Einstein immediately began to shape one of the new clips into a tool for rectifying the bent clip. Straus was bewildered, and asked Einstein what he was doing. The reply was, "Once I am set on a goal, it becomes difficult to deflect me."

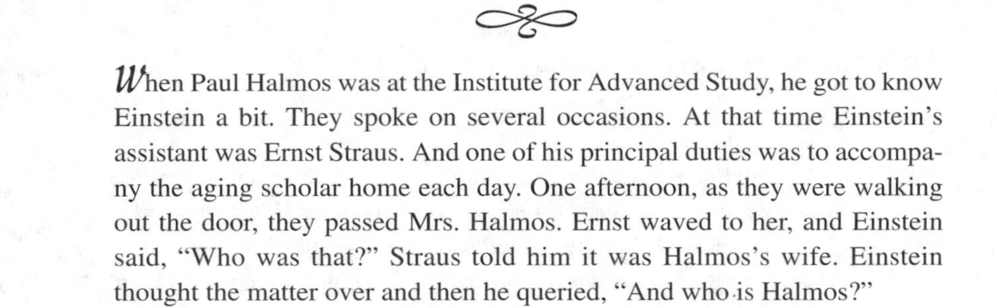

When Paul Halmos was at the Institute for Advanced Study, he got to know Einstein a bit. They spoke on several occasions. At that time Einstein's assistant was Ernst Straus. And one of his principal duties was to accompany the aging scholar home each day. One afternoon, as they were walking out the door, they passed Mrs. Halmos. Ernst waved to her, and Einstein said, "Who was that?" Straus told him it was Halmos's wife. Einstein thought the matter over and then he queried, "And who is Halmos?"

Albert Einstein lived in Princeton, a small New Jersey town that contains both Princeton University and the Institute for Advanced Study. Princeton police were trained to be on the watch for "eggheads" who would wander the streets aimlessly, or drive carelessly, or indulge in other unpredictable behavior. Einstein, who lived at 112 Mercer Street, one day took a wrong turn on the way home and became confused and lost. He wandered down Mercer Street, found the police station, and walked inside. He asked the officer at the desk whether he could be directed to Professor Einstein's house.

Joe Kohn (1932–) of Princeton was Stefan Bergman's assistant at Stanford in the late 1950's. Kohn had to listen each morning from 9:00 a.m. to noon to Bergman talking about his latest ideas.

One day, during the excitement of the Sputnik launching, Bergman came to Kohn and declared, "I think that the American people are ready to learn about complex 2-space. Go to the local television station and arrange for three hours in each of the next four weeks." Kohn was able to weasel out of this task, but still talks about it.

One of Bergman's proudest creations was the "Bergman kernel." Bergman wrote many papers and books on the subject. In his later years, whenever anyone proved something about the Bergman kernel, he was invited to Bergman's house for dinner. Bergman would strictly instruct the guest that, after dinner, he was to give a lecture to Bergman's wife Edy about the importance of the Bergman kernel.

At the end of World War II, Bergman found himself in France without any papers. This was always tricky, but at the time it was a matter of life and death: without papers, one could not obtain a rations card. Characteristically, Bergman found a mathematical solution to the problem. He went to the mayor of a small town outside Paris and convinced him to give Bergman a piece of paper saying, "This is to certify that Mr. Stefan Bergman has no papers." This of course was a paper, and with it Bergman was able to obtain his rations card.

It is said that, shortly after Bergman and his mistress arrived in the U. S., he took her aside and told her, "Now we are in the United States, where customs are different. When we are with other people, you should call me 'Stefan.' But at home you should continue to call me 'Professor Bergman.'" Others who knew Bergman well say that he was not the sort of man who would have had a mistress, that the man in question was more likely Richard von Mises (1883–1953), Bergman's sponsor. There is general agreement that the woman in question was Hilda Geiringer (1893–1973). In fact another story holds that Norbert Wiener went to D. C. Spencer (1912–2001) around this time and said, "I think that we should call the FBI (Federal Bureau of Investigation)." Puzzled, Spencer asked why. "Because von Mises has a mistress," was the serious reply.

In 1939 Hans Rademacher (1892–1969) owned an automobile, even though neither he nor his wife knew how to drive. He would have a friend drive them around from time to time. At one point Ivan Niven (1915–1999) offered to teach Rademacher to drive. After giving the matter some thought, Rademacher suggested that Niven teach Mrs. Rademacher. After all, she was in charge of the household, and he needed large blocks of time for uninterrupted concentration on mathematics. So Niven taught Mrs. Rademacher to drive; Professor Rademacher never did learn.

One day Norbert Wiener was walking across the MIT campus when someone stopped him with a question on Fourier analysis. Wiener pulled out a

Ron Graham

slip of paper and wrote out the answer in some detail. The interlocutor was most grateful, thanked Wiener, and began to go on his way. “Just one moment,” said Wiener. “Which way was I walking when we met?” The man pointed in the direction that Wiener was headed. “Good,” said Wiener. “Then I’ve had my lunch.”

In his later years, Paul Erdős (1913–1996) was addicted to benzedrine (uppers). He took them night and day. Ron Graham (1935–), who cared for Erdős in his declining years, worried that this substance abuse would affect Erdős’s health. He offered to pay Erdős a substantial sum of money if Erdős would refrain from taking his uppers for one month. Erdős assented, went through a month of abstinence, and collected the bounty. But he said afterwards that it was the most miserable month of his life. Each day he would spend the whole time staring at a blank sheet of paper, with no ideas in his head. He was greatly relieved when, at the end of the month, he was able to return to his addiction.

I wrote a paper with Erdős, so my Erdős number is one.[1] I also hosted Erdős at Penn State, where he gave two lectures. One day we took him to lunch. He ordered a fruit smoothie to drink, and it was served (as is customary) with a straw inserted and the straw paper left over the exposed end of the straw. Erdős made a huge show of trying to drink the beverage through the paper. At the end of the meal, the waitress asked us if we wanted coffee. Everyone said “yes” except Erdős. He said, “I have something much better than coffee,” and then he made a big production of bringing out his benzedrine.

---

[1] According to Casper Goffman, “And what is your Erdős number?” *Amer. Math.Monthly* 76(1969), 791, your Erdős number is zero if you are Paul Erdős. Your Erdős number is $\leq n$ if you have written a paper with somebody whose Erdős number is $n - 1$. Goffman’s *Monthly* article was answered by Erdős, “On the fundamental problem of mathematics”, *Amer. Math. Monthly* 79 (1972), 149–150, who expressed his delight with the concept and asked the following question: if each mathematician is a node on the face of the earth, and if any two mathematicians who have collaborated are connected with an edge, then is the resulting graph $\mathcal{G}$ planar? The answer “no” is also provided by Erdős: the graph $\mathcal{G}$ contains a copy of $\mathcal{R}(3,3)$, the complete bipartite graph on six vertices. That is, there are three vertices of each color black and white, and there are nine edges connecting black to white in all possible ways. Such a graph is non-imbeddable. The whites in this case are S. Chowla, Mahler, and Schinzel. The blacks are Davenport, Erdős, and Lewis.

Erdős's lecture at Penn State was a large public event. He was the first speaker in a lecture series intended for the entire campus. At the time, Erdős was at the height of his public fame/notoriety. His talk attracted 2000 people. I of course gave him a flowery introduction, telling of how I had carried all of Erdős's worldly possessions in a flight bag in one hand. The distinguished scholar walked, indeed limped, to the front of the room and immediately took off his shoe and told how his foot was sore and had been bothering him for weeks. A couple of weeks after Erdős left, my phone rang and I picked it up to hear (in a long, deep, heavily accented drawl), "Erdős. My foot is much better."

People used to say that you are not a real mathematician if you don't know Paul Erdős.

On April 8, 1974, Hank Aaron swatted his 715th home run. The prior record of 714 was held by Babe Ruth. These facts are a matter of great interest to sports fans and, to some extent, to the public at large. Carol Nelson (1951–), David Penney (1938–), and Carl Pomerance (1944–) of the University of Georgia found other fascination with these numbers. Namely, they noted that 714×715 is the product of the first seven prime numbers! It is also the case that the sum of the prime factors of 714 is 29, and the sum of the prime factors of 715 is the same. There are other pairs with this last property (15, 16 is one such), and these three mathematicians used a computer to generate many more. They published their results in the *Journal of Recreational Mathematics*, where they made a conjecture about the asymptotic sparsity of these so-called "Aaron pairs."

Soon after, Paul Erdős phoned Pomerance and told him that he could prove the conjecture. If Pomerance would pay his trip to Georgia, they could write a paper together. Erdős and Pomerance ended up writing that paper and more than forty more. Years later, both Hank Aaron and Paul Erdős were together at the Emory University to receive honorary degrees. Pomerance got the two to autograph a baseball together, so it can be said that Hank Aaron has an Erdős number of 1.

Of course David Hilbert was one of the gods of twentieth century mathematics (he also happens to be my mathematical great great great grandfa-

Photo courtesy Ron and Madelyn Gould.

1996 reception at Emory University. Ron Gould (standing), Gary Hank (on right), Paul Erdős, and Hank Aaron.

ther). One day Hilbert and his wife were entertaining some people for dinner. About half an hour before the guests were to arrive, Mrs. Hilbert told David to go change his clothes. He went upstairs, took off his jacket, his shirt, his shoes, and his pants. Being a creature of habit, he then brushed his teeth, got into bed, and went to sleep. When the guests arrived, Hilbert was nowhere to be found. Mrs. Hilbert had to go upstairs and awaken her husband. [This is a popular and fondly told story—one which Mrs. Hilbert herself took pains to deny.]

On another occasion, the Hilberts went to someone else's house for dinner. When they got home, Mrs. Hilbert chided David for not wearing a necktie. He subsequently mailed a necktie to the people, and instructed them to stare at it for three hours.

Helmut Hasse (1898–1979) told the following story in a letter to Constance Reid:

> During the first post-war congress of the Gesellschaft Deutscher Naturforscher und Ärtze held in Leipzig, when we met in the evenings in the Burgkeller there was much questioning of the type, "What about Professor *X* from *A*, is he still alive?" On

> such an occasion I happened to be seated together with other young people quite near to the table where Hilbert was seated together with the cream of the participants. I heard him put exactly the above type of question to a Hungarian mathematician about another Hungarian mathematician. The former began to answer, "Yes, he teaches at *B* now, he occupies himself with the theory of ..., he was married a few years ago, there are three children, the oldest ..." But Hilbert already, during the first few words, began to say repeatedly, "Ja, aber ...," "Ja, aber ...," and when he eventually succeeded in stopping the other fellow's flow of words, he continued, "Ja, aber, das wollte ich ja gar nicht alles wissen. Ich habe doch nur gefragt, 'existiert er noch?' " ["Yes, but I did not wish to know all that detail. I only asked 'Does he still exist?' "]

A few years later, Hilbert uttered his famous statement (with reference to Brouwer's program), "To deprive a mathematician of existence proofs would be like depriving a boxer of his gloves."

Fuld Hall is the main building of the Institute for Advanced Study in Princeton. Old and austere, it is built in the classic Georgian style. There is a rather formal room, off the entrance, where afternoon teas are held. A painting in the tea room depicts an old man shying away from an angel coming down from the clouds. In the old days, Stanislaw Ulam (1909–1984) used to joke that this painting depicted Mina Rees (1902–1997)—then the director of the Office of Naval Research—proposing a Navy contract to Einstein, who is recoiling in horror.

When the Institute for Advanced Study was new, it was temporarily domiciled in the center of the Princeton University campus. John von Neumann, a founding member of the Institute and a celebrated character, was famous for the parties he gave; late in the night one could see liquor bottles sailing out of the windows. Eventually the university administration put a stop to that.

One day von Neumann was walking past the elegant Gothic Princeton University Chapel. He said, "This is our one-million-dollar protest against materialism." Once, caught in a traffic jam with Ulam in the rain, he said, "Mr. Ulam, cars are no good for transportation anymore, but they make marvelous umbrellas."

Stan Ulam accompanied Norbert Wiener to meet G. H. Hardy at the train in South Boston when he came to MIT for a visit. They brought along Norman Levinson (1912–1975), and afterward the group went (at Wiener's invitation) to a Chinese restaurant for lunch. Immediately Wiener started speaking Chinese to the waiter, who seemed not to understand a word. Wiener remarked, "He must be from the south and does not speak Mandarin." After lunch, Wiener picked up the check but discovered that he had no money. A collection was taken and the lunch paid for. Wiener scrupulously reimbursed everyone later.

G. H. Hardy was once riding on a train in Britain. Sitting across from him was a schoolboy reading an elementary algebra book. Endeavoring to be friendly, Hardy asked the lad what he was reading. "It's advanced mathematics," came the reply. "You wouldn't understand."

John E. Littlewood had a "niece" named Ann Streatfeild. He frequently vacationed with her and was evidently quite fond of her. She called him "Uncle John." In the 1970's he became concerned that she was not married and, after he was gone, there would be nobody to care for her. It turned out that Ann Streatfeild was Littlewood's daughter. He kept the matter a close secret, but the secrecy bothered him. Eventually Béla Bollobás and his wife convinced Littlewood to tell people that Ann was his daughter. He did so one night in the Combination Room at Cambridge, simply by speaking of his "daughter" rather than his niece. He was chagrined that, next day, nobody had evidently taken any notice of this change. Eventually Ann Streatfeild married Carl Johannsen of Zurich, and Littlewood was very happy indeed.

The preceding story seems to remove all doubt about J. E. Littlewood's sexual orientation. On the other hand, Littlewood was once asked about Hardy's. He said, "Hardy is a non-practicing homosexual."

G. H. Hardy once scrawled a five-letter, schoolboy-level term of contempt on a pupil's paper. On second thought he decided this was a bad idea and endeavored to erase the word. He was unable to do so. So he told the student that the paper was lost. The pupil was both pious and hyperconscientious; he kept clamoring for the return of the paper, to no avail.

Littlewood relates that the Cambridge Common Rooms are full of people talking about the decadence of Oxford; and the Oxford Common Rooms are full of people talking about the decadence of Cambridge. Littlewood himself was once sitting in the Cambridge Commons drinking port, and observed that his neighbor was an Oxford man. By way of explanation he said, "It certainly wasn't your manner; must be an inferiority complex on my part." The Oxford man took him to be in complete earnest, and spent considerable time trying to cheer Littlewood up.

In the 1950's at MIT there was a stellar group of young mathematicians who created an atmosphere of creativity and excitement. These included Paul J. Cohen (1924–), E. M. Stein (1931–), Adriano Garsia (1928–), Alberto Calderón (1920–1998), Jürgen Moser (1928–1999), and John Nash (1928–). They would have spirited discussions, challenging each other to solve problems and to formulate theorems. These chalkboard debates would spill over into the lunch period, and there was much mathematical activity at the math lunch table in Walker Hall. Norbert Wiener liked to join the young mathematicians and to bask in their admiration. One day Cohen turned to Wiener and asked in mock candor, "Professor Wiener, what would you do if one day when you went home you were to find Professor *X* sitting on your living room sofa?" Cohen was alluding to a well-known mathematician who was reputed to indulge in the dubious practice of nostrification (Hilbert's term for the Göttingen habit of ignoring other mathematicians' contributions in favor of one's own). Wiener became red in the face and retorted, "I would throw him out and start counting the silver!"

Stories about Marcel Riesz's (1886–1969) unworldliness and impracticality abound. It is said that he could not pack a suitcase, nor could he figure out how a mailbox worked. He had the habit of hiding his paychecks under the mattress instead of cashing them. Adriano Garsia had to run to banks all over town cashing checks for Riesz the day before Riesz was scheduled to take one of his trips to Sweden.

Kurt Gödel's move from Vienna to the United States was abrupt but necessary for his survival. In March of 1939, the Nazis abolished university lectureships and created a new position called *Dozent neuer Ordnung* (Lecturer of the New Order). There were no such positions for Jews. Gödel was out of a job, and he also received a letter ordering him to report to the German army for a physical examination. This was a man who had recently spent quite a lot of time in sanatoria for overwork and depression. So Gödel was out of a job, and facing the prospect of becoming a grunt in the army.

When Gödel next got a bill from his cleaning lady, he found that it listed the total due (which was *Reichsmark 6.80*) and then, underneath, a neatly typed *Heil Hitler!*. For Gödel this was the last straw. He decided that he and his new wife (Adele Nimbursky, the nightclub dancer) had to move to Princeton (where he had an offer). They decided that traveling by boat across the Atlantic was too risky, so they took the long way around—across Russia, the Pacific, and the continent of North America. It can truly be said that, for Gödel, the Institute was (in Abraham Flexner's words) "a paradise, a haven where scholars and scientists could regard the world and its phenomena as their laboratory, without being carried off into the maelstrom of the immediate."

On one occasion, Gödel attended one of the Institute's twice-yearly formal dinners. He was seated across from John Bahcall (1934–), the famous young astrophysicist. The two introduced themselves, and Bahcall related that he was a physicist. Gödel replied scornfully, "I don't believe in natural science."

John von Neumann always dressed like a banker, in a formal suit with necktie and pocket handkerchief. There were various Freudian theories about why he felt the need for this pageantry, but there you have it. Once von Neumann and his wife went on a vacation to Arizona, where they visited the Grand Canyon. Of course von Neumann wanted to see everything, and have the full experience. So they arranged to take a trip to the bottom, and were to ride down on pack mules. All the other participants showed up in the expected dress—short sleeves, chaps, cowboy boots, sombreros, and the like. John von Neumann had on the usual banker's suit, necktie, and pocket square.

Von Neumann also had various "absent-minded professor" qualities. Once his wife, when she was ill, sent him to get her a glass of water. He came back after a while and asked where the glasses were kept (of course he could blame this situation on the servants). They had only lived in that house for seventeen years. On another occasion, von Neumann drove out from Princeton in the morning to meet an appointment in New York City. But halfway there he forgot whom he was supposed to meet. He phoned his wife and asked her, "Why am I going to New York?"

In the early 1980's, Erdős became a cult figure in the popular press. There were extensive interviews with him, some quite revealing. For example, in one such he was asked why he never had married. The obvious answer, which the interviewer probably expected, is that Erdős had spent most of his adult life living with his mother, and furthermore that he had no resi-

Paul Erdős

dence and no regular source of income. Instead Erdős said, "This may sound very strange to you, but I cannot tolerate sexual pleasure."

Erdős liked to joke about death, and about senility. He frequently made jokes about the loss of brain power. He said that there are three steps in the mental degeneration of a mathematician:

- First you forget your theorems.
- Next you forget to zip up.
- Last you forget to zip down.

During World War II, top secret work was being conducted at Los Alamos (code name Shangri La), Oak Ridge (code name Dogpatch), the Chicago "Metallurgical Laboratory," and other sites as well. One of the main goals was the development of the atomic bomb. It really was classified. But in fact there were a few hundred physicists and mathematicians who were aware of it. Paul Erdős, who in some sense was an "enemy alien," really worried a group of physicists when he sat down to lunch one day and asked, "The bomb based on atomic fission that you're building—is it ready yet?"

Paul Halmos was of course thrilled when Marcel Riesz visited the mathematics department at the University of Chicago. Not only was Riesz a most eminent mathematician, but he was a fellow Hungarian. Halmos was at that time a tenure-track member of the department—one of the boys—and he scurried over to pay his respects. Speaking Hungarian, he introduced himself and welcomed the celebrated mathematician to Chicago. Riesz replied, "Glad to see you, sonny," and then he immediately began dictating a letter. What could Halmos do? He wrote it down.

R H Bing (1914–1986) was a topologist at the University of Texas, a student of R. L. Moore (1882–1974). Many a Texas gentleman will forsake his first and middle names and use only his initials. Bing in fact *did not have* a first and middle name. His full name, as recorded on his birth certificate, was R H Bing. These were *not* initials, and there were no periods. One year

Bing applied for a visa so that he could take a trip abroad. He filled out the application form, giving his name as usual. The form was returned to him, and he was told that the State Department would not accept initials. The applicant must give complete names. Bing wrote back, explaining patiently that his name was " 'R only' 'H only' Bing." In a few weeks he received his visa in the mail. It was made out to "Ronly Honly Bing."

William "Bus" Jaco was a student of R H Bing at the University of Wisconsin (before Bing went to Texas) in the 1960's. In those days Bing had a great many students, and he once organized them in vans to go to a mathematical event in Michigan. When they got there someone posed a math question that he had been saving up for the visiting topologists. Jaco was quite taken with the question. He skipped the reception, skipped dinner, and stayed up all night thinking about the problem. He was determined to solve it and show Wisconsin proud. At 6:00 am the following morning he had not solved it, and he went somewhat dejectedly off to breakfast. The only other person he found in the room was Bing himself, and Jaco plopped down next to his advisor with an air of defeat. Bing asked him why he was up so early, and Jaco confessed that he had attacked this topology question, which he related on the spot to Bing. Bing stared off into space for a while, scratched his head, then grabbed up a napkin and wrote out a beautiful and elegant solution. Jaco was overwhelmed. He knew that the famous R H Bing was a great topologist, but this was just too much. He told Bing how impressed he was. "Well," said Bing, "I, too, heard the problem yesterday. And I, too, skipped dinner and the reception and stayed up all night thinking about it. I just got it half an hour ago."

R H Bing was fond of saying that when he was young he would rather *give* a math talk than listen to one. "But," he went on, "now that I am older I would rather give *two* math talks than listen to one.

The Mittag-Leffler Institute in Djürsholm, Sweden has a fine library, with many unusual books. On the third floor, in the corner, looking out over "his books," is an enormous bronze statue of Gösta Mittag-Leffler (1846–1927) himself (commissioned by same) holding his reading glasses. In 1972,

Gösta Mittag-Leffler

Dennis Hejhal's (1948–) collaboration with Max Schiffer on the kernel function led him to discover in Mittag-Leffler's library a marvelous book by H. F. Baker (1866–1956) called *Abel's Theorem and the Allied Theory, Including the Theory of the Theta Functions*, published by Cambridge University Press in 1897. Later, Lars Ahlfors (1907–1996) was visiting the Mittag-Leffler Institute when Hejhal wrote to him describing his results and particularly praising the book of Baker. Hejhal offered to come to Mittag-Leffler if the master complex analyst wanted to hear more. Ahlfors replied that he'd rather wait until his return to Harvard (Hejhal was in Minnesota at the time), and went on to say that he found Baker's book, "but that in order to get the book, I first had to get a ladder, prop it up against the statue, and quite literally stand *on top of* Mittag-Leffler to grab it from the (nearly inaccessible) shelf above his head." The whole uncanny situation reminds one of Isaac Newton (1642–1727) standing on the shoulders of giants.

The book collection at the Mittag-Leffler Institute is not large, but it has many rare books and manuscripts. In fact John Garnett (1941–) found an unpublished MS of Otto Frostman (1907–1977) that contained results that have proved to be quite influential in the work of Garnett and his students Peter Jones (1952–) and Don Marshall (1947–). On one occasion, a visitor from France came to spend time at the Mittag-Leffler Institute. Ostensibly his purpose was to study mathematics, but in fact he had been commissioned by his university to steal a particular book. Unfortunately, he succeeded.

My friend Glenn Schober (1938–1991) was once teaching a class to help train graduate students to teach. For the first day he carefully crafted a lecture on elementary mathematics in which he purposely made 25 cardinal teaching errors. He walked into class on that first day, told the students he was going to give a sample lecture, and did so. At the end he said, "There were a number of important teaching errors that I intentionally committed during this lecture. See how many you can identify." The students found 32 errors.

Do you have trouble remembering the quotient rule? When you write it down for your freshman calculus class, do you sometimes get the numerator backwards? Then do what I do: use the "*Quotient Rule Song*," written by Edward Newman for Robert A. Bonic's *Freshman Calculus* (I was a co-author of this book). It goes as follows (and you can provide your own tune):

If the quotient rule you wish to know,
It's low-de-high less high-de-low.
Then draw the line and down below,
Denominator squared will go.

What else can one say except, perhaps, Q.E.D.?

In the old days at the mathematics institute in Göttingen, it was the tradition for each new faculty member to dress in tails and a top hat and to call on the senior professors at their homes. One of these fresh young fellows called on the Hilberts. Mrs. Hilbert admitted him, and he sat down and deposited his hat on the floor. Hilbert was distracted with mathematics, and had no patience for this courtly nonsense. He listened irritably to the polite social chatter until he could stand no more. Then he took the top hat from the floor, placed it on his own head, grabbed his wife's arm, and said, "My dear, I think we have delayed our good colleague long enough." Then Dr. and Mrs. Hilbert abandoned their own house.

The following story is generally told about either Bertrand Russell (1872–1970) or A. N. Whitehead (1861–1947). The subject once claimed that, given $1 + 1 = 1$, he could prove any other statement. One day some wise guy said to him, "OK. Prove that you are the Pope." The great sage thought for a while and then proclaimed, "I am one. The Pope is one. Therefore the Pope and I are one."

Denis Diderot (1713–1784) once paid a visit to the Russian court, at the invitation of the Empress. He was allowed to converse freely, and he gave vent to his many thoughts on atheism. The Empress was amused, but some of her courtiers were not. She was advised to put a muzzle on some of Diderot's doctrines. On the other hand, she did not like to curb her guests' tongues. So the following ruse was devised. Diderot was informed that a certain learned mathematician had an algebraic proof of the existence of God, and that he was willing to present it to the court. Diderot consented to this demonstration.

At the appointed hour, Leonhard Euler (1707–1783) came to court. He advanced toward Diderot and said in a very serious tone, with perfect conviction, "Monsieur, $(a + b^n)/n = x$, therefore God exists. Any answer to that?" Unfortunately, poor Diderot was completely ignorant of mathematics. He was embarrassed and at a loss. He asked permission of the Empress to return to France at once, and it was granted.[2]

[2] It may be noted that scholars have investigated this story in detail. There is considerable evidence that it is apocryphal—in the strong sense of the word.

Edwin Beschler and Joseph P. LaSalle

One day R H Bing was driving a group of mathematicians to a conference. As usual, Bing launched into a detailed discussion of a particular problem in topology that he wanted to solve. The passengers were nearly as interested in the math as was Bing, but they were rather nervous because the weather was bad, visibility poor, and there was lots of traffic. Bing did not seem to be giving the matter his full attention. To make matters worse, the windshield was completely fogged over and it was nearly impossible to see. A sigh of relief was quietly breathed when Bing began to reach toward the windshield. Everyone supposed that he was going to wipe the windshield clear and pay attention to what was happening on the dark and stormy road. Instead Bing started drawing diagrams with his finger in the windshield mist.

Mathematics publisher Edwin Beschler, creator of the esteemed Academic Press Pure and Applied Mathematics series, once had occasion to repeat to Harley Flanders an anecdote he had been told by Janos Aczel. The substance was that Aczel had been in a restaurant, ordered his dinner, and then observed the waiter to "disappear with all his derivatives." Flanders looked quizzically at the publisher for a few seconds, then said, “Aczel must have said ‘VANISHED with all his derivatives.’” Only then did he laugh.

Great Affrontery

Norbert Wiener gave a lecture in Göttingen. Afterward, David Hilbert stood up and said that the tradition in Germany was to give bad lectures. He went on to say that Göttingen had the worst reputation among institutions in Germany, and that the mathematics department had the worst reputation in the university. Wiener kept puffing himself up, anticipating a great compliment. Then Hilbert said, "But Wiener's lecture is the worst that I have ever heard."

Wiener was near-sighted and peculiar in a number of ways. He was walking down the street one day in Germany in the 1920's, wandering all over the place. He nearly ran into a pedestrian coming the other direction; indeed, the fellow had to step off the sidewalk so that he could pass Wiener. The man was a German military officer and proud. He handed Wiener his card and challenged him to a duel. His second would call on Wiener in due course. Wiener was of course extremely upset and consulted a German friend. The friend said he would be Wiener's second, and he would handle the matter. When the officer's second called, Wiener's friend pointed out that, because Wiener was the respondent in the duel, he got to choose the weapons. He went on to say that Wiener wanted to choose weapons indigenous to his native land: tomahawk and hatchet at three paces. The challenge to a duel was quietly withdrawn.

Perhaps the least delightful arena in which we all wrestle with standards is that of referee's reports. The *Annals of Mathematics*, Princeton's journal, has very high standards and exhorts its referees to be tough-minded.

George Pólya

Solomon Lefschetz (1884–1972) was instrumental in establishing the preeminence of the *Annals*. But I doubt that even he could have anticipated the following event. Many years ago, Gerhard Hochschild (1915–), who sets high standards for himself and for everyone else, submitted a paper to the *Annals*. The referee's report said, "Good enough for the *Annals*. Not good enough for Hochschild. Rejected."

There are many versions of the story of how George Pólya (1887–1985) was forced to resign his post at Göttingen. Here is how Pólya himself related the tale in a letter to Ludwig Bieberbach (1886–1982):

> The story of the boxed ear is completely childish and, by the way, it goes like this: On Christmas 1913 I traveled by train from Zürich to Frankfurt/M and at that time I had a verbal

> exchange—about my basket that had fallen down—with a young man who sat across from me in the train compartment. I was in an overexcited state of mind and I provoked him. When he did not respond to my provocation, I boxed his ear. Later on it turned out that the young man was the son of a certain *Geheimrat*; he was a student, of all things, in Göttingen. After some misunderstandings, I was given a *consilium abeundi* (i.e., told to go away) by the Senate of the University. (The story, even privately, is not worth defending.)

Pólya decided not to contest the case, and he abandoned Göttingen. [In some versions of the story Runge's father-in-law, a lawyer, endeavored to defend Pólya—to no avail.] He went almost immediately to Paris. He ultimately found his way to Stanford, where he had an illustrious career.

Pierre-Simon Laplace (1749–1827) presented a copy of his celebrated *Mécanique Céleste* to Napoleon Bonaparte. The great leader studied the book assiduously. He then sent for Laplace and declared, "You have written a large book about the universe without once mentioning the author of the universe."

Laplace replied, "Sire, I have no need of that hypothesis."

Joseph Louis Lagrange (1736–1813), who worked with Laplace on the book of mechanics, later observed that, "It is a beautiful hypothesis just the same. It explains so many things."

Not very long ago, certainly within many of our lifetimes, it was quite difficult for female mathematicians to get academic jobs. Emmy Noether (1882–1935) came to Bryn Mawr in the U.S.A. because she could not get a proper professorship in Germany. Guido Weiss (1928–) had many outstanding offers in the mid-1960's, but he came to Washington University because it was the *only* university to offer his wife Mary (1930–1966) a job comparable to the one they were offering to him. [The university that came in second offered the pair a single job that they could "split up any way they wanted."] Mary Ellen Rudin (1924–), a distinguished professor with an endowed chair at the University of Wisconsin and a member of the National Academy of Sciences, recounts that she had a sequence of "non-jobs" until 1971, when she became a full professor at the University of Wisconsin. She

Mary Ellen Rudin

looks at the matter philosophically, and points out that one of the benefits of having a non-job is that she could teach whatever she liked, she did not have to serve on committees, and she never had to teach trigonometry. She was able to have graduate students, apply for grants, travel, and be a mathematician. She has no complaints.

There are many tales of G. H. Hardy and John E. Littlewood, and of their collaboration. It is said that when Wiener first met Littlewood he exclaimed, “Oh, so you really exist. I thought that ‘Littlewood’ was a name that Hardy put on his weaker papers.” It should be noted that there are many variations of this story, some involving Landau instead of Wiener. In fact here is one such version:

It is said that Landau thought that "Littlewood" was a pseudonym for Hardy (so that it would not seem that Hardy was writing all the papers). Landau visited Cambridge, never saw Littlewood, and returned to Göttingen convinced that his theory was correct.

When Besicovitch met Hardy, he is quoted as saying, "I know you. Who are you?"

Besicovitch, a Russian by birth, was imbued with nineteenth century traditions. After leaving Russia (a prudent move on account of his rumored black market dealings during World War I), Besicovitch ended up at Cambridge University in England. A dinner was given in his honor, at which the main course was a delicious game bird. In his thick Russian accent, Besicovitch asked the name of the tasty food that they were eating. When he heard the reply, he exclaimed, "In Russia we are not allowed to eat the peasants."

Besicovitch was a quick study, and he rapidly became proficient at English. But his fluency was never perfect, and he always had a notable accent. He adhered to the Russian paradigm of never using articles before nouns. One day, during his lecture, the class chuckled at his fractured English. Besicovitch turned to the audience and said, "Gentlemen, there are fifty million Englishmen speak English you speak; there are two hundred million Russians speak English I speak." The chuckling ceased.

The book *Invarianten Theorie* by Roland Weitzenböck (1885–1955), written in 1923, has a remarkable feature. The "Vorwort" (Foreword) of the book has the property that if you take the first letter of each sentence in sequence, then it spells out "Nieder mit den Französen," which means "Down with the French." A similar story can be told about my former teacher Bob Bonic (1932–1990). While he was putting the finishing touches on his book *Linear Functional Analysis,* he wanted to exhibit his ardor for his new girlfriend. But he was disinclined to reveal this ardor to his then wife. He solved the problem as follows. If one looks at the first letter of each paragraph in the "Remarks and References" to the book (Chapter VII, page 116, ff.), one can see the name "Joanna Pang" spelled out. In order to get it right, one must note that the last relevant paragraph begins with "And now, going in a different direction ..." The act of going in a different direction gives the needed "ang."

Mitchell Taibleson (1929–) was a student of E. M. Stein (also my teacher) in the early 1960's. His thesis—a real *magnum opus*—laid the foundations of the theory of Lipschitz spaces. Stein thought that this was a seminal piece of work, and advised Taibleson to publish it all together in a single, prominent journal. So Mitch submitted it to *Acta Mathematica*, one of the pre-eminent journals in analysis. The journal happens to be Swedish. Unfortunately, at that same time the Swedes were developing their own theory of Lipschitz spaces. In fact Jaak Peetre (1935–) was the ringleader, and he was doing things from the point of view of Besov spaces. One of the Swedes (most likely Hörmander (1931–)) refereed Taibleson's paper and rejected it. Since this was Taibleson's first publishing effort, he was somewhat abashed. He went to Antoni Zygmund (1900–1992)—Stein's teacher—for advice. Zygmund said, "Break it up into ten papers and send it to ten different journals. Your reputation will be established immediately. You can worry about being ethical later."

One day two Cambridge mathematics professors were discussing an impending policy change in the math department. Finally, out of some frustration, the first savant said, "But we've been doing it this way for the past 400 years!" "Quite so," said the second professor, "but don't you think that the last 400 years have been rather exceptional?"

Pythagoras (c. 569 B.C.–475 B.C.) one day saw a puppy being beaten. He took great pity on the poor beast and said, "Stop, do not beat it; it is the soul of a friend which I recognized when I heard it crying out."

Robin Wilson (1943–) tells of the day that Aneurin Bevan (1897–1960), the politician who founded the British health service in the Attlee government, asked his father, Prime Minister Harold Wilson (1916–1995), what young Robin was going to be when he grew up. Mr. Wilson said, "Well it

Robin Wilson

looks as though he's going to be a mathematician." Bevan replied, "Just like his father, all bloody facts, no bloody vision!"

Bertrand Russell once asked Littlewood what was the French word for a certain English tree. Littlewood's rejoinder was that he did not see the use of knowing the French word for a thing that he would not know in English. Russell's reply was, "Not at all; Bevan knows the word for 'woman' in at least 14 languages, but he wouldn't know one if he saw one."

The FBI file on Richard Courant (1888–1972), obtainable through the Freedom of Information Act, contains the following peculiar statement:

> Reports of investigations dated 31 March, 1943 and 27 May, 1943, Boston, Massachusetts, stated that Courant was regarded by two professors at Massachusetts Institute of Technology, Cambridge, Massachusetts, as a brilliant scientist, but one who was tricky and unscrupulous in plagiarizing the works of other men. For that reason, they felt he should not have access to confidential or secret information. They stated that Courant's first loyalty was to himself; that he was troublesome in his dealings with others; and because of his many associations with the German race, it was possible he might inadvertently blurt out some secret in the presence of an enemy agent.

It should be noted, however, that Courant ultimately did receive a security clearance.

Mathematicians are not always charitable in their characterizations of their colleagues. A. N. Kolmogorov (1903–1987) once said that brilliant mathematicians peak young. For example, he pointed out, I. M. Vinogradov (1891–1983), peaked at age 4, pulling legs off of insects. Vladimir Arnol'd (1937–) peaked at age 15, and he (Kolmogorov) peaked at 8.

Peter Duren once gave a lecture at the Mittag-Leffler Institute in Djürsholm, Sweden. Arne Beurling (1905–1986) was in the audience, and he asserted that he had proved Duren's result years before, but had not published it (Beurling was famous for this sort of thing). Duren boldly told Beurling that that was quite interesting, but that he (Duren) intended to publish the result anyway. For the rest of the afternoon, one could see Beurling chasing Duren around the grounds.

Peter Duren was once dandled on Albert Einstein's knee—when his father was a visitor at the Institute for Advanced Study.

Richard Feynman (1918–1988) was a rather mathematical sort of physicist, winner of the Nobel Prize. His second marriage, which by his own telling

A. N. Kolmogorov

was done "on the rebound," did not work out well. It was decided that a divorce was best for all concerned. Following the custom (and the law) of the time, one partner had to file charges against the other. Among civilized folk, the usual pattern was for the woman to charge against the husband. Mrs. Feynman's accusation was "He gets up in the morning and immediately starts to do calculus. And in the evening he plays his bongo drums." End of marriage.

Don Sarason (1933–) is one of Paul Halmos's most distinguished students. When Sarason was a graduate student, Halmos invited him over for dinner. Sarason attended, and a good time was had by all. The next day, he dropped by Halmos's house and left off a large and fancy box of candy for Mrs. Halmos. She was out at the time, but Halmos said he would give it to her. A few days later, Mrs. Halmos asked Paul whether he had told Sarason how much she had enjoyed the candy. He had not, so the next day he sought out Sarason and said, "My wife says I'm a slob—I should have thanked you for the candy." "That's all right," answered Sarason, "my mother told me to do it."

Today, Sarason has a very casual manner. He is usually clad in blue jeans and a T-shirt, and he wears his long grey hair in a ponytail. But he has a large collection of neckties on a rack in his office, and he usually drapes one around his neck when he goes off to teach. Sarason is the son of a former accountant for General Motors. His uncle was a bigwig—actually Treasurer of GM. In the late 1960's, Sarason drove a fancy late model Chevrolet that was purchased from a GM executive through a special incentive program. From this (perhaps) grew the story that Don Sarason, as a student, always drove a late model GM car.

Sarason says that the following is his favorite Halmos story. When Halmos was young, and AMS meetings were much more intimate, Halmos (being a gregarious type) would approach strangers and say, "Hi, I'm Paul Halmos. Who are you?" At one meeting he did this to Ken Hoffman (1930–) , and about half an hour later he did it again to Hoffman. Hoffman found this irritating. After consulting with John Wermer (1927–), he decided to retaliate. He approached Halmos and said, "Hi, I'm Ken Hoffman. Who are you?" A few years later, Halmos was invited to give a colloquium talk at MIT (Hoffman's institution). At the appointed hour, Hoffman entered the tea room where Halmos and several locals were assembled. When he spotted Hoffman, Halmos rushed over and said, "Hi, you're Ken Hoffman. Who am I?"

Paul Halmos is from Hungary, and sometimes takes pleasure in reminiscing about the Iron Curtain mentality. He says that, in the old days in Russia, if you could read and write, then you were called intelligent. If you could read or write but not both, then you were called a specialist. In Hungary, cops cruised in twos: one who could read and one who could write. In Romania, however, they cruised in threes: one who could read, one who could write, and one who kept an eye on those untrustworthy members of the intelligentsia.

Paul J. Cohen (Fields Medalist, 1966), who has been a professor at Stanford University for many years, always liked to complain about how lazy the students were. One day he fell asleep in his office. In fact he slept through the class he was supposed to teach. Then he went to the office of his neighbor to complain about how lazy the students are. He followed this by returning to his office and going back to sleep.

André Weil was famous for writing tart and pungent letters of recommendation. It got so that people were afraid to ask him for a letter. But one confident young fellow did so. The letter read, in part, "This man is better than anyone you now have, anyone you ever have had, or anyone you ever will have." The man got the job.

André Weil was fond of saying that there was a logarithmic law that governed hiring. In brief, first-rate departments hire first-rate people, second-rate departments hire third-rate people, and third-rate departments hire fifth-rate people. The logical conclusion of this analysis is that the university administration must not allow complete self-determination to second-rate (or lower) departments. It must intervene to keep them on the right course.

André Weil, spiritual leader of the Institute for Advanced Study for many years, set a high standard. In 1973, Associate Professor Michael S. Mahoney of the History of Science Department at Princeton University had the temerity (or perhaps the bad luck) to write a biography of Pierre de Fermat (1601–1665). At that time, Weil had been studying the history of Fermat for some time. He had given a lecture series on the subject. The depth of his understanding was uncanny: Weil had actually figured out sequences of letters that had been sent, on what dates they were mailed and had arrived, and who was thinking what when. He fancied himself to be the pre-eminent Fermat scholar. And the new biography by his colleague down the road did not strike his fancy. Somehow it was arranged for Weil to review the book for the *Bulletin of the AMS*. Weil begins the published review by reminding us that "In order to write even a tolerably good book about Fermat, a modicum of abilities is required." He then lists these prerequisites:

- Ordinary accuracy.
- The ability to express simple ideas in plain English.
- Some knowledge of French.
- Some knowledge of Latin.
- Some historical sense.
- Some familiarity with the work of Fermat's contemporaries and of Fermat's own mathematics.
- Knowledge of and sensitivity to mathematics.

André Weil then proceeds to consider each of these attributes one by one, and to demonstrate—via annotated quotations from the book under review—that the author apparently possesses none of them.

André Weil was nothing if not irreverent. He once described the Taj Mahal as "the bastardized offspring of Italian baroque grafted onto the ostentatious whims of a Mughal despot." When Weil was told that a certain mathematician had proposed a certain theorem, Weil dismissed the subject out of hand. He said, "That can't be true. Because if it were true, *he* wouldn't know it."

In the mid-1980's, Weil gave a lecture series at the Institute for Advanced Study on the subject of Pell's equation (John Pell, 1611–1685). He pointed out that the equation is named after an English mathematician who really had nothing to do with it. Properly speaking, the equation should be named after Fermat. Nevertheless, Weil allowed, he would follow common usage and call it "Pell's equation." "This has happened many times in mathematics," Weil went on to explain in accented English. "For example, I live on von Neumann Circle. I live there… but it is *still* called von Neumann Circle." With a shrug and a barely perceptible twinkle in his eye, he returned to the mathematics.

Olga Taussky-Todd (1906–1995) relates that Ernst Zermelo (1871–1953) was a frustrated and resentful man. When he attended the same conference with Kurt Gödel, Zermelo made a point of telling people that he did not want to meet the famous logician.

It happens that a luncheon was planned one day at an inn on top of a small mountain. A group planned to hike up the mountain and then to dine. Zermelo was in the group, and some of his friends thought that this would be a perfect opportunity for him to meet Gödel. Zermelo immediately bridled. He mistook somebody else in the group for Gödel, and said that he could not speak to somebody with such a stupid face. This misunderstanding was explained away, but then Zermelo said that there would not be enough food if Gödel were invited along. When that did not work, Zermelo said that the hike was too arduous for him; he could not make his way up the mountain. Finally, someone simply brought Gödel up to Zermelo and introduced him.

At that point, says Taussky-Todd, a small miracle occurred. Zermelo and Gödel almost immediately became lost in a deep conversation about logic. The two walked effortlessly up the mountain without even knowing that they had done so.

At one of the pre-eminent math departments in the Midwest, one fairly senior faculty member was singled out for promotion to full Professor. His was not a spectacular record of achievement, but he had been a solid worker for many years and he had contributed steadily, both to mathematics and to the department. He was getting on in years, and it was felt that this encomium was long overdue. So the case was prepared, and outside letters solicited. One letter, from a French mathematician, read in part as follows: "Perhaps he could be a Professor in the provinces, but *never* in Paris." The case, unfortunately, died.

In graduate school, I used to play chess with (the now famous physicist) Frank Wilczek (1951–). One day Frank, then a second year graduate student in mathematics, remarked over a chess game that he had just solved the thesis problem that D. C. Spencer had given him two weeks earlier. I was immediately jealous, but fascinated to hear more. So I asked how Spencer reacted to this news. Frank said, "Spencer was delighted. He told me to write it up and he'd get me a job at MIT." Now I was even more jealous. So I asked how Frank had responded. He said, "I told Spencer that I wouldn't put my name on this sh**. I'm going over to the physics department and write a real thesis." Which he did.

Later, Wilczek made quite a name for himself while at the physics institute in Santa Barbara. One day he got "the call" from the Institute for Advanced Study. They made him a powerful offer, with a fine salary and many other perks. After the negotiations were nearly complete, Wilczek said that there was just one more thing he required: He wanted to live in Einstein's house. If the Institute would buy him the house at 112 Mercer Street, he would accept the job. They did and he did.

Heinz Hopf (1894–1971) spent his later years at Indiana University. One day there was a faculty meeting to consider the hiring of a particular candidate for a faculty position. Considerable discussion ensued, and one point that was repeatedly raised was that this person would be stimulating for Hopf. Certainly Hopf would enjoy having him around. At the end of this

Heinz Hopf

discourse, the assembled faculty were still undecided—as there were other candidates—and they turned to Hopf for an opinion. Hopf scowled and said, "Keep that man away from me."

As we all do, Hopf finally became rather aged and his lightning cerebrations slowed down a bit. At one point he went up to someone in the math department and said, "I hear that the great Professor Hopf is visiting us next week."

I had the privilege, during one of my many visits to Indiana University, to meet Max Zorn (1906–1993)—the namesake of Zorn's Lemma. Zorn was quite old at the time, probably in his late eighties. He still came into school most days, went to tea, hung out with the mathematicians, and so forth. Unfortunately, not long after my visit, Zorn was walking to school one day and was mowed down by a student rushing to class in his car. Zorn survived, but he contracted pneumonia in the hospital and then died.

On another visit to Indiana, I had a great deal of trouble finding a place to park my car, and found my way to the math building after everyone had gone to lunch. It happened to be Secretary's Day, so all the staff was gone as well. As I wandered the halls forlornly, trying to decide what to do, a gentleman ran up to me with one of my books in his hand. He cried, "You're Krantz! Please autograph my book." Of course I was delighted, and proceeded to take out my pen. Meanwhile, he opened his book to the flyleaf for my convenience. He then frowned and said, "Oh damn. You have already signed it."

Stefan Bergman thought intensely about mathematics and cared passionately about his work. One day, during the International Congress of Mathematicians in Cambridge, Massachusetts, Bergman had a luncheon date with two Italian friends. Right on schedule they appeared at Bergman's office: the distinguished elder Italian mathematician Mauro Picone (1885–1977), bearing a bouquet of flowers for Bergman, and his younger colleague Gaetano Fichera (1922–1996). This was Picone's first visit to the United States, and he spoke no English; Fichera acted as interpreter. After greetings were exchanged, Bergman asked Fichera whether he had read Bergman's latest paper. Fichera allowed that he had, and that he thought it was very interesting. However, he said that he felt that the existence of the Green's and Neumann functions in this context had not been established. Bergman said, "No, no, you don't understand," and proceeded to explain on the blackboard. Picone, comprehending none of this, waited patiently. After the explanation, Bergman asked Fichera whether he now understood. Fichera said that he did, but he still thought that there was some issue about *existence*. Bergman became adamant and a heated argument ensued—Picone comprehending none of it. After some time, Fichera said, "Well, let's forget it and go to lunch." Bergman cried, "No existence—no lunch!" and he remained in his office while the two Italians went to lunch. Picone gave the flowers to the waitress.

At some point Fichera's wife acquired a dog; the question arose as to what to name the beast. Alexander Weinstein told Fichera that the dog should be called "Courant." Fichera resented the suggestion, as he was good friends with Richard Courant.

It is said that, in the 1930's, Marston Morse (1892–1977) and his wife were not getting along. Mrs. (Celeste) Morse ended up consulting Professor William Fogg Osgood (1864–1943) of Harvard about her marital problems. To make a long story short, Osgood divorced his German wife and married Celeste Morse, who had meanwhile divorced her husband. There was a meeting of the National Academy which both Osgood and Morse attended; people wondered whether the two scholars had traveled together and discussed "our wife."

Olga Taussky-Todd, who edited Hilbert's collected number-theoretic works, was having tea in Brouwer's house, engaged in an animated discussion about Hilbert. L. E. J. Brouwer (1881–1966) said he thought some of Hilbert's papers were not entirely his own; he made reference to the famous walks with Hermann Minkowski (1864–1909), during which many ideas were exchanged. Brouwer said there was only one paper of Hilbert which he was sure was Hilbert's own: the solution of Waring's problem, for he wrote this when he was a guest in Brouwer's house.

Olga Taussky-Todd

When Stefan Bergman was at Brown, one of Bergman's graduate students got married. The student planned to attend a conference on the West Coast, so he and his new bride decided to take a bus to California as a sort of makeshift honeymoon. There was a method in their madness: the student knew that Bergman would attend the conference, but that he liked to get where he was going in a hurry. The bus was too slow, and seemed to be the least likely mode of transportation for Bergman. But when Bergman heard about the impending bus trip, he thought it a charming idea and purchased a bus ticket for himself. The student protested that this bus trip was to be a part of his honeymoon, and that he could not talk mathematics on the bus. Bergman promised to behave.

When the bus departed, Bergman was at the back of the bus and, just to be safe, Bergman's student took a window seat near the front with his wife in the adjacent aisle seat. But after a half hour, Bergman got a great idea, wandered up the aisle, leaned across the scowling bride, and began to discuss mathematics. It wasn't long before the wife was in the back of the bus and Bergman next to his student—and so it remained for the rest of the bus trip. The story has a happy ending: the couple is still married, has a son who became a famous mathematician, and several grandchildren.

Witold Hurewicz (1904–1956) was a brilliant mathematician, noted for his work in dimension theory, differential equations, and many other fields as well. Once Hurewicz and Lefschetz attended a conference in Mexico. In some sense they were rivals: they competed in solving many of the same problems. The two of them took an afternoon off and climbed one of the pyramids. Only Lefschetz came down (in the usual way). He said that Hurewicz was so overcome by the view that he stepped back and fell to his death. Certain misanthropes later suggested that Lefschetz may have given him some help.

Charles Loewner (1893–1968) was a complex analyst whose ideas played a seminal role in Louis de Branges's proof of the Bieberbach Conjecture. When Loewner first came to this country, he landed a job at the University of Louisville. His students quickly realized that Loewner (who soon moved to Stanford by way of Syracuse) knew a great deal more mathematics than anyone else on the faculty. So they asked him to spend some extra hours with them, teaching them advanced mathematics. Loewner agreed, and they met regularly before class. When the chairman of the mathematics department found out about this arrangement, he put a stop to it. He felt that this activity was taking Loewner's time away from teaching.

My friend Richard Arens (1919–2000) could be disarmingly frank, especially when he was in his cups. One evening, at a social gathering, I asked him how old he was. He said, "Biologically I'm 56. But I was such an idiot for the first 18 years of my life that I like to say that I'm 38."

One evening when Arens was feeling particularly puckish, he came up to me and said, "Young man, do you want to be famous?" I was a young

Paul Erdős and Charles Loewner

Assistant Professor at the time, and there was hardly anything I wanted more. So I said, "Yes, indeed I do." Arens said, "Well, then fu** up a subject."

One day a colloquium speaker at UCLA referred several times to a book by the author S— (a living author), calling it "The second worst mathematics book ever written." The temptation was too great, so one of us after the talk asked him about this sobriquet. He said, "S— has written another book."

Hardy liked to joke about his friends and collaborators. One such was George Pólya. Pólya once had a good idea, one of which Hardy approved. Afterwards, Pólya (by his own telling) did not work sufficiently hard to develop the thought, and he did not bring it to fruition. Hardy was not happy. Later, Hardy visited a zoo in Sweden with Marcel Riesz. In a cage there was a bear. The cage had a gate, and on the gate was of course a lock. The bear sniffed at the lock, hit it with his paw, then he growled a bit, turned around, and walked away. Hardy said, "He is like Pólya. He has excellent ideas, but he does not carry them out."

G. H. Hardy once told his friend Bertrand Russell that if he could find a proof that Russell would die in five minutes, he would naturally be sorry to lose him—but the sorrow would be quite outweighed by the pleasure in the proof. Russell, quick to understand the ways of mathematicians, said, "I entirely sympathized with him and was not at all offended."

A friend once encountered Bertrand Russell in a state of deep meditation. Asking what occupied him so, he heard in reply, "Because I've made an odd discovery." The great philosopher looked rather weary. "Every time I talk to a savant I feel quite sure that happiness is no longer a possibility. Yet when I talk with my gardener, I'm convinced of the opposite."

Sergei Nikitovich Mergelyan (1928–) has a celebrated theorem to his name—one that can be found in many textbooks. Namely, in 1950 he proved a theorem about polynomial approximation that was much better than anything that any of the experts thought could possibly be true. In 1990, I visited Cornell and was pleased to discover that Mergelyan was there at the same time. I told my hosts that I wanted to meet the great man. Thus, later in the day, we were brought together. Mergelyan, ever gracious, said to me, "Oh, Professor Krantz. I am so pleased to meet you." I reciprocated that I was honored to meet Professor Mergelyan. Then he said, "And I am so sorry that I have missed your talk." My hosts chimed in and said, "Oh, no. You haven't missed his talk. In fact it is at 2:00 p.m." Then Mergelyan said, "Then I am sorry that I am *going* to miss your talk."

Eugene Wigner (1902–1995) was a distinguished mathematical physicist, winner of the Nobel Prize. In the late 1940's, after World War II, Wigner was one of many distinguished participants in a mathematical physics seminar that was held regularly at the University of Chicago. On a particular day, the speaker was a young man who knew all the latest jargon. He was speaking about hydrodynamics, but he used vector bundles and differential forms and characteristic classes and exact sequences and he had almost everyone baffled. At the end of the talk, the eminent Wigner stood up to pose a question. Wigner was well known to be quiet and self-effacing. He said, "Excuse me. I need to pose a question. I am not quite sure how to put it. It's just that… Well, let me put it this way. I mean to say… I mean… I want to ask… I hope you don't mind my putting it… Ahem, I think that… What I want to ask is: Where does the water go?"

Bertrand Russell was once asked whether he was prepared to die for his beliefs. "Of course not," he replied immediately. "After all, I may be wrong."

An American mathematician of some note was returning from a trip abroad and had to go through Customs. The U. S. Customs Officer asked him what he had been doing during his one-week sojourn. The reply was that he had been at a mathematics conference. The Customs Officer then took this man aside and detained him for some time with a great many tedious questions about exactly where he had been and what he had been doing during his travels. The mathematician kept glancing nervously at his watch, worried that he would miss his connecting flight. The Customs Officer finally got to a point of asking our friend what he had had for dinner each day. Finally the mathematician threw up his hands and exclaimed, "Why are you doing this to me?" The Customs Officer smiled and said, "Ah. Now you know how I felt when I took calculus."

Littlewood once encountered a mental block in that he could not remember the name of Lord Cherwell, which was Ferdinand Lindemann. He adopted the mnemonic that it was the same name as that of the man who proved that

Ferdinand Lindemann

π was transcendental (Ferdinand von Lindemann, 1852–1939). The end result was that Littlewood could not remember *his* name either.

A great tradition from the old days at Cambridge University is the Mathematics Tripos Exam, which is an oral exam taken by every student after a prolonged and agonizing period of study. In Littlewood's day, the Mathematics Tripos was a three-year course, though sometimes students would take the exam at the end of the second year. It was four days of tests of up to ten hours per day, and these were tests of speed, of knowledge, and of accuracy. According to Béla Bollobás, the Tripos was, in its prime, the most severe mathematical test the world has ever known. There is no analog for it today. The Exam evolved during the eighteenth century. From 1753 onward, the examinees were divided into three classes: Wranglers,

Senior Optimes, and Optimes. Wranglers were the best. John E. Littlewood was a Senior Wrangler, bracketed with a man named Mercer. In fact Senior Wranglers were celebrities in Cambridge, and their photographs were sold during May Week. A friend of Littlewood's endeavored to buy his photo, but was told, "I'm afraid we're sold out of Mr. Littlewood [who was in fact a dashing and handsome fellow] but we have plenty of Mr. Mercer."

Hardy opposed the Tripos, declaring that it was nonsensical. In point of fact, the Tripos never seemed to produce many good pure mathematicians, but it was successful at training applied mathematicians. As a demonstration of the foolishness of the exam, Hardy persuaded George Pólya to take the mathematics Tripos without previous coaching. To Hardy's chagrin, Pólya got a high pass, and in future years his friends enjoyed calling him a Senior Wrangler. In 1910 Hardy played a decisive role in abolishing the strict order of merit of the Tripos. He was also a sworn enemy of the milder exam which replaced it. Littlewood also firmly opposed the order of merit. He felt that he had wasted his first two years at Cambridge studying for the exam.

Littlewood once made the critical remark that, "*X* knows the right thing to say about the label on a wine bottle, but didn't know a corked bottle when he met it. The image of the English Tripos."

Leopold Infeld was a docent in Lwow (also, for historical reasons, spelled Lviv, Lemberg, Lvov, Leopoli, L'wiw, L'vov, and Lwiw), Poland in the 1930's. After a month's stay in England, he gave the following characterization of British intellectual conversations versus Polish intellectual conversations. In Poland, he said, people talked foolishly about important things; but in England they spoke intelligently about foolish things.

Stan Ulam spent twenty years of his career at Los Alamos, part of the time (during the war years) developing the atomic and hydrogen bombs, and part of the time working on other scientific projects. Obviously some of this work was classified, and some of it very highly classified. In the fall of 1957, Ulam was a visiting professor at MIT. He was assigned an office across the hall from Norbert Wiener. One day Wiener stopped Ulam in the hall to say, "Ulam! I can't tell you what I'm working on now, you are in a position to put a secret stamp on it!"

John von Neumann was once in a physics lecture in Princeton. The lecturer exhibited a slide with many pieces of experimental data and, although they were badly scattered, he argued that most of them lay on a curve. It is said that von Neumann murmured, "At least they lie on a plane."

Stan Ulam was never noted for his modesty, and his autobiography *Adventures of a Mathematician* amply displays his ego. His wife once chided him for lack of humility. He said, "True. My faults are infinite, but modesty prevents me from mentioning them all."

The physicist Enrico Fermi (1901–1954) was quite knowledgeable about mathematics and spent much time with mathematicians. But he loved to tease everyone. He used to enjoy taunting Edward Teller (1908–) with the question, "Edward-a how com-a the Hungarians have not-a invented anything?" He needled Emilio Segré (1905–1989), an ardent fisherman, when Segré was expounding on the difficulty of catching trout. Fermi said, "Oh, I see, Emilio, it is a battle of wits." Ulam was once marveling at how accurately the Schrödinger equation predicted the levels of energy in a hydrogen atom. Fermi said, "It has no business being that good, you know, Stan."

Noting that the first proof of any new result is often clumsy and inefficient, Abram Besicovitch once said that "A mathematician's reputation rests on the number of bad proofs he has given."

On reading a paper that he did not much like, Besicovitch said, "The surprising thing about this paper is that a man who *could* write it—would.

The analyst Edmund Landau kept a printed form in his office for handling "proofs" of Fermat's last theorem that came in over the transom. It read "On page___ , lines___ to ___ , you will find that there is a mistake." Actually *finding* the error was a task that fell to a Privat Dozent.

Mathematical notation is extremely important. It not only lends clarity and precision to the writing of mathematics, but it shapes the way that the reader sees the subject. Camille Jordan (1838–1922) had his own view of the matter. It was said that if he had four objects of equal weight (such as variables a, b, c, d), he would call them a, M_3', ε_2, and $\Pi_{1,2}''$.

Littlewood believed that eight hours a day of work was more than sufficient for a mathematician. By contrast, Landau had an enormous capacity for work, often putting in 12-hour days. He was also a gluttonous eater; as a result, he was in the habit of taking an 80-minute siesta after lunch. Normally Landau did not exercise, but he took long walks—sometimes up to 20 miles. He worked to completely rigorous rules. Once Littlewood and Landau were working together in Cambridge. They began immediately after breakfast. At some point Littlewood said, "Excuse me for a minute or two." When he returned, Landau said, "Two minutes, 47 seconds."

Bertrand Russell and G. E. Moore (1873–1958) were once having a philosophical discussion. At one point Russell said, "You don't like me, Moore, do you?" Moore replied, "No." That point having been disposed of, the conversation resumed.

Littlewood recounts that after a period of very easy relations, a chill came between himself and Alexander Ostrowsky (1893–1986). He asked his colleague Besicovitch whether he had had a similar experience. Besicovitch replied that, indeed, he had. Littlewood asked whether he knew why. Besicovitch allowed that he did. Asked for an explanation, Besicovitch told Littlewood this story: "Well, Ostrowsky and I were taking two ladies to the cinema. Ostrowsky's lady asked him, 'Professor Ostrowsky, what is your subject exactly?' He replied, 'You should ask Professor Besicovitch that.' I said, 'The fact is that I have never read any of Ostrowsky's work.' "

Emmy Noether

Once Landau was asked to affirm that Emmy Noether was a great woman mathematician. He replied, "I can testify that she is a great mathematician, but that she is a woman, I cannot swear." (Some later photographs suggest that Noether was a very plain woman, although her more youthful portraits show that she was quite attractive. Both Paul Aleksandroff and Einstein described her in glowing terms. Alexandroff spoke poignantly of her femininity.)

Bertrand Russell and J. E. Littlewood each met Einstein once—at a ceremonial dinner. Sitting near Einstein was a sporting peer who got quite drunk. He heckled Einstein across the table—in English of course. Einstein knew no English at the time, and could not understand a word of what was being said. The peer boasted later that he had stumped Einstein on relativity. Russell claimed (to Littlewood) that all he remembered of the evening was that

Einstein told him a dirty story (in German). Littlewood replied that Einstein was well known for consummate tact in adapting himself to his company.

One of the Einstein legends concerns an event just after his arrival at the Institute for Advanced Study in Princeton. He went for a walk on the idyllic grounds, and he met J. Robert Oppenheimer (1904–1967) on a footbridge. At that time Einstein knew no English and Oppie knew no German. Of course they each knew who the other was, and they wanted to be friendly; but both were at a loss. After a bit, Oppie pointed down at the water and said, "Fish." "Ja," said Einstein, "Fisch."

Ralph Boas (1912–1992) was not only a fine mathematician but he was also an exemplary citizen. He served in many capacities for the AMS and the MAA, and was a terrific writer and editor. On top of all that, he was the chairman of the math department at Northwestern University for many years. According to his son, Harold, Ralph would come into work each morning, climb the fire escape, and enter through the window to his chairman's office. That way he would not have to see anyone and he could get right down to work.

E. Harrison once asked Bertrand Russell, "Is it true that philosophy has never proved that something exists? Russell replied, "Yes, and the evidence for it is purely empirical."

Once, at the Mathematical Sciences Research Institute, Reinhold Remmert (1930–) approached me and said that, in recent years, he had been studying mathematical history. In particular, he had spent much time on Bieberbach. "You know," he said, "Bieberbach was a terrible anti-Semite and bigot. He used to have his office a few doors down from mine, and he was a dreadful person—completely irascible and unpleasant." He then told me that he was due to give a talk on Bieberbach the following week. He had heard that Bieberbach had some unusual middle names. Could I research the matter on the Internet for him (this was in 1995, the dawn of the Internet age)? With the help of a friend, I conducted the necessary research. Most historical sources give Bieberbach's full name as Ludwig Bieberbach.

Ludwig Bieberbach

However there is a Scottish Web site at St. Andrews on the history of mathematics, and it provides the full name of the notorious anti-Semite. It is

Ludwig Georg Elias Moses Bieberbach

The next day, after I gave Remmert the information he required, he was quite grateful and voluble and wanted to say some nice things to me. He told me how much he liked my book *Function Theory of Several Complex Variables*. I thanked him for the nice compliment. Then he looked off into the distance and said, "Of course Karl Stein (1913–2001) hates the book."

Bieberbach wrote a dreadful book called *Aryan Tractat of Mathematics*. In it, he claimed that Gauss achieved his great insights through pure Aryan genius, while Carl Gustav Jacobi (1804–1851) got his from "dirty Jewish tricks."

In 1939 Albert Einstein and his friend Léo Szilárd (1894–1964) wrote to the President of the United States advising him that "extremely powerful bombs of a new type" could be made from fissionable materials such as

uranium. The physicists asked that the President "speed up the experimental work" which was then being conducted. Einstein was quite willing to do more than letter-writing to help in the war effort. Thus it was that, in December of 1941, Vannevar Bush (1890–1974), Director of the Office of Scientific Research and Development, asked for Einstein's help with certain problems arising from the gaseous diffusion method of separating U-235 from other uranium isotopes. Einstein was happy to help, and sent Bush a handwritten manuscript with copious technical information. He said that he would do more, but that he required additional scientific data to do so.

Bush would not provide Einstein with the necessary information. The famous physicist was considered to be a security risk. Bush said, "I wish very much that I could place the whole thing before him and take him fully into confidence, but this is utterly impossible in view of the attitude of people here in Washington who have studied into his whole history."

As a result, Einstein spent the war years in Fuld Hall at the Institute, vainly seeking the unified field theory that would elude him for the rest of his life.

Everybody knows that the Fields Medal is the highest encomium that can be bestowed on a mathematician. Established with the last will and testament of J. C. Fields, it is designed to recognize the work of a "young" mathematician. By custom, but not by mandate, the cutoff age for the Fields Medal is 40 years of age.

Traditionally, the criteria for getting a Fields Medal were of a very cut and dried nature: you state very important theorems and you prove them. Lars Valerian Ahlfors received one of the first Fields Medals for brilliant results in the geometric function theory of one complex variable. Jesse Douglas (1897–1965) also received one of the first Fields Medals for solving the Plateau Problem. Pierre Deligne (1944–) received the Fields Medal for proving the Weil conjectures. Paul J. Cohen received the Fields Medal for proving the independence of the continuum hypothesis. And so forth. In recent times, the criteria for the Fields Medal have been liberalized. Some have received the Fields Medal for creative thought, not always backed up by proofs. A few mathematical physicists have received the Medal. One wag characterized the situation as follows:

- (William P.) Thurston (1946–) showed us that you can get the Fields Medal without proving a theorem.
- (Edward) Witten (1951–) showed us that you can get the Fields Medal without stating a theorem.

The old Fine Hall at Princeton, one of the world's most celebrated mathematics buildings, was special in part because it was designed by the mathematicians themselves. There was a beautiful, covered garden walkway, so that the faculty could have their thoughtful peregrinations. The building had lovely paneled offices and beautiful, leaded-glass windows with mathematical formulae and portraits. The old Fine Hall even had showers. The irreverent young folk in the Fine Hall common room added these verses to the classic Princeton undergraduate song about the faculty:

Here's to Veblen, Oswald V.,
lover of England and her tea;
he built a country club for math,
where you can even take a bath.

A pair of alternative last two lines for Veblen were

He's the one mathematician of note
who needs four buttons to button his coat

The explanation is that Veblen wore distinctive coats: long, straight, with four buttons.

Solomon Lefschetz also had a verse in the faculty song. It was

Here's to Lefschetz, Solomon L.,
unpredictable as hell;
when he's laid beneath the sod,
he'll start right in to heckle God.

Herbert Robbins (1915–2001), famous for many reasons but in particular for the wonderful book *What is Mathematics?* that he co-authored with Richard Courant, was chairman of the mathematics department at Columbia University. One day he was vigorously defending a tenure case, but the Dean was skeptical. He didn't think the poor fellow's research program had any momentum. He was worried that, in a few years, it would dry up. "Don't worry," said Robbins. "If his research program dies, then we'll make him a Dean."

It is not well known that Richard Courant did not want Robbins's name on the book *What is Mathematics?*—even though Robbins did a substantial portion of the writing. Even today, the copyright is held only by Courant's son, not by Robbins or his heirs. Robbins was never contractually entitled to any of the royalties, although Courant did send him a sum (at his own discretion) from time to time. These payments were not perpetuated by Courant's son.

In 1984 I spent a month in Sweden. I was lucky enough to be invited to the annual meeting of the Scandinavian Mathematical Society. It was held in Lund, and I went with much anticipation. Jaak Peetre, whom I had never met, agreed to pick me up at the airport. He said he would hold up a copy of *Math Reviews* so I would know who he was. It worked like a charm.

One of the ceremonies at these meetings is that two honorary fellowships are awarded to promising young mathematicians. This is a great encomium, and quite a fuss is made. In 1984 the honorees were Svante Janssen and Anders Melin (1943–). Part of the pageant is that, to help appreciate each young honoree, some famous mathematician gives a short talk about his work. Peetre gave the talk about Janssen's work. He said that Janssen was extremely prolific, he was only two years from the Ph.D. and had already written 50 papers, Peetre hadn't read many of these papers and wasn't sure he understood any of them… . A bit compulsive, but it certainly conveyed a very positive impression of Janssen and served the purpose nicely. Then Lars Hörmander (Fields Medalist, 1962) got up to give the presentation about Melin's work. He began by saying, "Anders Melin has written only two papers. But they are *good*." It should be explained at this point that Lars Gårding was in the audience at all the talks. He was Hörmander's thesis advisor. He was in a feisty mood at this meeting, and he had been heckling all of the speakers—even speakers about whose work he knew nothing. So when Hörmander began the mathematical portion of his remarks by saying, "The story of Melin's work begins with Gårding's inequality," of course Gårding perked up. What Hörmander actually wrote down was the Fefferman–Phong inequality, which is a spectacular generalization of Gårding's venerable inequality. It is notable for having a Sobolev term of negative order on one side. Gårding immediately piped up and said, "That's *not* my inequality!" Hörmander smiled and replied, "This is *much better* than your inequality." Gårding was blessedly silent after that.

It is said that, when Hörmander wrote his thesis with Gårding, he was held to a very high standard. Gårding made his student re-write and re-write—

to the point that some of the other faculty began to wonder what was going on. At one point someone asked Gårding, "Why are you tormenting this young man so? He has a great mathematical career awaiting him. Why don't you let him get on with it?" The reply was, "This is the only thing he will ever write that I will understand. I want it to be just right."

When Lars Hörmander won the Fields Medal in 1962, the Swedes were, of course, thrilled. To this day, he is the only Swede to have garnered this honor. He was invited to a dinner of state with the King of Sweden, and was in fact seated next to the monarch. It should be noted that Hörmander is a fine man, very friendly, but he does not lack for self-assurance. At one juncture early in the evening the King of Sweden turned to Hörmander and asked, "Haven't we met before?" Without hesitation, Hörmander replied, "I don't remember."

Hörmander once told me of a mathematical trip that he took to the Soviet Union. When he got off the plane he realized that he had no hotel accommodation. He endeavored to get access to a telephone directory so that he could phone a hotel. But, in those days in the Soviet Union, all phone books were controlled by the state. Nobody would let him look at one. And nobody would look up the number for him. Hörmander also could not obtain the phone numbers of his hosts. So he flew *back* to Stockholm, got the phone numbers he needed, and flew again to Mother Russia.

Wilhelm Magnus (1907–1990) was great with students, and he directed many Ph.D. theses. Magnus allowed that he did not mind writing a student's thesis for him. But he did find it quite irritating when the student kept coming by to check on his progress.

Great Ideas

It used to bother Littlewood a good deal that he seemed to dream at night of solutions to the problems he was working on. But he could never remember the details in the morning. Littlewood found the situation to be more and more aggravating, and he resolved to address this predicament; he put a notebook and pencil by his bedside. That night, he had a particularly lucid dream in which all the pieces of the solution of his problem were plainly laid out and explained. When the dream occurred, he forced himself to wake up and he wrote down all his thoughts at the moment he was dreaming them. Then he went blissfully back to sleep. In the morning, he was excited to read what had been recorded, confident that he would now be able to write an important new paper. What he read on his tablet was, "Higamus, bigamus, men are polygamous. Hogamus, bogamus, wives are monogamous."

Paul Halmos is a lover of language. He wrote the book *How to Write Mathematics*, and it has had a strong influence over the shape and form of the subject. Once, in considering the issue of whether one should end a sentence with a preposition, Halmos constructed this sentence that ends with five prepositions:

> What did you want to bring that book that I didn't want to be read to out of up for?

[There is a version of this story set in a lighthouse, and in that context the quotation ends with "to out of up around for?"

In a lecture series on approximation theory, Besicovitch announced, "zere is no 't' in ze name Chebyshóv" (P. Chebychev, 1821–1894). Two weeks later he said, "Ve now introduce ze class of *T*-polynomials. They are called *T*-polynomials because '*T*' is ze first letter of ze name Chebyshóv."

Einstein used to say that "Insanity means we keep trying the same thing and hope it comes out differently."

My teacher J. J. Kohn was the man who cracked the $\bar{\partial}$-Neumann problem, and this was a major mathematical event. D. C. Spencer, C. B. Morrey, and others had thought about the problem in the 1950's but had come nowhere close to solving it. Almost nothing was known in 1961 when Kohn cracked it.

Obviously very excited at his triumph, Kohn phoned up Spencer on the morning of Easter Sunday and told him the news. Spencer's excitement exceeded Kohn's. He dropped everything and demanded that Kohn meet him at Fine Hall (the Princeton math building) and tell all.

Spencer raced into Princeton (in those days he lived out on Route 1, near the Forrestal plasma physics campus) and drove along Lake Carnegie. He ended up behind an extremely slow-moving car, could not get around it, and became quite frustrated. At one point, Spencer leaned his head out the window, pounded on the horn, and shouted, "Get off the road, you sons of bitches!"

Pretty mild by today's standards, and Spencer forgot all about it in the ensuing moments. He sped into the Princeton campus, spent several hours with Kohn, and relished the ingenious solution that Kohn unfolded for him.

Several days later Spencer received a summons from the Princeton magistrate. He was charged with "public obscenity," and was to appear in court on a specified day. That he did, and there was the driver of the other vehicle—a sedate young fellow with his wife and their baby.

The magistrate asked the young man what he had been doing when the "incident" occurred. The answer was that he was driving his wife and baby daughter to church for Easter Sunday. Then the magistrate asked Spencer what he had been doing. The answer was that he was driving into Fine Hall to hear about the solution of the $\bar{\partial}$-Neumann problem. At that point Spencer felt that he had already lost half the battle.

The judge then wanted to hear about the sordid details of the "incident." He admonished those present that he wanted no repetition of the obscenity in his courtroom. Spencer admits that he then made a tactical error: he giggled.

To make a long story short, D. C. Spencer lost the hearing. He was fined $25 (quite a bit of money in those days). But he got to be the first to hear about an important mathematical event.

Barry Simon (1946–) is a mathematical physicist at Cal Tech. He is noted for his prolific output of scientific work, including books, papers, and many writings on computers. He has even designed, and written a book about, his own operating system for the PC. He has a monthly column in a popular computer magazine.

Once Barry was on a trip to a large university which was out in a rural area—as most large public American universities are. So the last leg of his trip was on a small plane. Somewhere along the way the plane got into trouble—fire in one engine or some such thing—and an emergency landing took place. The passengers were quickly evacuated from the plane. Everyone was greatly relieved to see that all made it to safety, and there was not the slightest injury. But Barry Simon looked most unhappy. Someone asked him why and he said, "I lost five papers that I was working on." "Oh," the friend said, "but surely you kept copies back in your office." Barry's reply was, "Not of the two papers that I wrote on the plane."

And now some arcana of Washington University (my home institution). It is a little-known fact that Henri Poincaré (1854–1912) discovered special relativity theory at almost exactly the same time as Albert Einstein. According to Stephen Hawking, Poincaré's paper appeared just two weeks after Einstein's. Even less well known is the fact that Poincaré was invited to participate in the 1904 World's Fair (a centennial celebration of the Louisiana Purchase) that was held in St. Louis—in Forest Park, just across the street from the current location of the university. The heart of the campus today consists of venerable buildings that were originally constructed to be administration buildings for the Fair. One of these is Cupples I, which now houses the mathematics department. Another is Holmes Lounge, where Poincaré delivered a lecture on—guess what?—special relativity. And this was in 1904, one year before the appearance of Einstein's celebrated paper on the same topic. In fact a translation of Poincaré's lecture was published in a journal called *The Monist*, vol. 15, January, 1905. The title is "The Principles of Mathematical Physics." While the paper is largely philosophical, it is also the case that certain pages could have been lifted from a modern freshman physics text. It treats the Michelson–Morley experiment, discusses the standard topic of a train traveling at the speed of light with two observers, and many other familiar parts of special relativity theory.

It is moderately well-known that the St. Louis World's Fair introduced hot dogs and ice cream cones to the world (and later on the Washington University Medical School, adjacent to Forest Park where the Fair was held, introduced molecular biology and AIDS to the world), but now we see that the World's Fair introduced relativity to the world.

More recently, Holmes Lounge spawned the celebrated and successful 1978 movie *National Lampoon's Animal House*. Don't believe it? Read on. Harold Ramis was a student at Washington University. And Ramis was both the writer and producer of *Animal House*. He proudly recounts that he spent many an afternoon in Holmes drinking coffee with his friends and planning this movie. And Washington University certainly had a notorious Beta House (now defunct), the model for the fraternity in the movie. The Washington University Chancellor, William Danforth, would not allow the film to be *made* at Washington U. In fact, it was filmed at the University of Oregon.

Shizuo Kakutani is one of the great experts on Brownian motion. He understands better than most that Brownian motion in two dimensions is generically recurrent: motion beginning at a point P has positive probability of returning to P; but in three dimensions this is false. He likes to describe the situation by saying, "A drunken man will usually find his way home. A drunken bird has no hope." [This story is also attributed to George Pólya.]

I spent the academic year 1995–1996 at the Mathematical Sciences Research Institute in Berkeley. That was "the year of the Unabomber" (Theodore Kaczynski, 1943–). Bill Thurston, then the Director of MSRI, had been conscripted by the FBI to read the Unabomber's long polemic that was published in the newspapers and to help to identify him. Of course Bill could not talk about the matter until it was all over, but he told us then that it was immediately clear to him that the author of this essay was a mathematician.

After the Unabomber was caught (he was in fact turned in by his own brother), the FBI took great pains to interview people who had known him. One of these was Reese Harvey (1941–) of Rice University, who had been an instructor at U. C. Berkeley at the same time as the Unabomber. Reese showed up at his office in shabby clothing with several days' growth of beard, only to be confronted by a pair of FBI agents. He said, "Oh, I've just come from my isolated cabin in the woods. I have no phone, no plumbing, and no electricity. I could not be reached." This may have rung a bell for the agents.

I was driving down the street one day in Berkeley, listening to the radio, when they began interviewing John Addison (1930–), who had been chairman of the math department during the Unabomber's time in Berkeley. He said that he had known all along that the Unabomber was someone from the Berkeley math department. He just hadn't known that it was *this particular guy*.

The 1995–1996 academic year was a bad one for mathematics. That same year, Walter Petryshyn (1929–) of Rutgers administered thirty blows with a hammer to his wife's skull, and thus murdered her. Seems that he never got over some of the errors in his recent book *Generalized Topological Degrees and Semilinear Equations*, published by Cambridge University Press.

Dennis Sullivan tells the story of R H Bing and Saunders MacLane (1909–) meeting in the late 1940's on a train going to the annual AMS meeting in January. Of course they were both topologists, MacLane from the elegant and distinguished Harvard tradition and Bing from the more down-to-earth Texas tradition of R. L. Moore. Both men were members of the National Academy of Sciences. The conversation went as follows:

SM: Hello, RH.
RHB: Howdy, Saunders.
SM: Say, RH, what have you been working on?
RHB: I've been working on the pseudo-arc.
SM: The pseudo-arc? Say, I've never heard of the pseudo-arc.
RHB: Well, Saunders, if you've never heard of the pseudo-arc, then the "plane continua that you have heard of" form a set of first category in the space of all plane continua.

Saunders MacLane was a remarkable young man. He graduated from Harvard and went on to the University of Chicago for his graduate studies. He tells me that, soon after he arrived in the windy city, he realized that the Chicago math department had declined considerably. It seems that the august faculty had been replacing themselves, upon retirement, by their own students. So he packed up and moved to Germany, where he studied with Paul Bernays (1888–1977) and Hermann Weyl (1885–1955). One evening MacLane found himself during the intermission at the opera standing next to Hitler and Goebbels! He has sometimes mused that he wished he'd had a gun.

When publisher William Jovanovich (1920–2001) was a student at Harvard, he often dined at a cafeteria that was famous for its cheap but dreadful food. Bertrand Russell also ate there. Of course Russell came from a quite distinguished and wealthy British family. One day Jovanovich could no longer restrain his curiosity. He went up to Russell and said, "Mr. Russell, I know why I eat here. It is because I am poor; but why do you eat here?" Came the reply, "Because I am never interrupted."

Hans Heilbronn (1908–1975) was a number theorist. He was a student of Edmund Landau in Göttingen. One of his colleagues at Cambridge, Patrick Duval, professed to be "half afraid of him." Heilbronn was a formidable mathematician by any measure. In 1918 Hecke showed that if one assumes an extended version of the Riemann hypothesis, then one can prove that $h(d) \to \infty$ as $d \to -\infty$, where $h(d)$ is the class-number of the quadratic number field of discriminant d. Heilbronn astonished the mathematical world by proving in 1934 that if one *denies* the same generalized form of the Riemann hypothesis then the same conclusion obtains. It follows (*tertium non datur*) that the conclusion holds in any case.

I had a friend who was a geophysicist. He was also a math groupie, and we had many a stirring discussion over a cold beer of mathematics and geophysics and their similarities and differences. One day he got a very serious look on his face and said, "You mathematicians. You think ideas are important, don't you?" I allowed that, yes, ideas had their use in mathematics. He went on to say, "And you value originality, don't you?" I said, yes, originality was important to us. He sat up straight and proclaimed, "Well, they are not important to geophysicists! Those original guys with ideas just confuse us. What is important in geophysics is organizing data." I had never before entirely understood the meaning of the word "speechless."

My friend Ed Dunne (1958–), now at the AMS, was a graduate student at Harvard. One semester he was a TA, and part of his duties was to write certain exams. One of the exams was in a subject with which Ed was not entirely conversant. So he went to the math library, where a large book is kept that contains old exams. Ed paged through it looking for ideas. Along the way, he came across an old algebraic geometry exam written by expert David Mumford (1937–). Dunne was suitably awed. Mumford is a great genius, holder of the Fields Medal and the MacArthur Prize, and admired by all. Ed paused to see what sort of algebraic geometry exam Mumford might write. It had just two questions. They read:

1. Write an exam for this course.

2. Take it.

It is rumored that the Harvard philosophy department has a similar exam. It has a third question:

3. Grade it.

Ed Dunne

There are a number of stories circulating about mathematicians who were famous for giving terse exams. One such exam at MIT is supposed to have said, "You have a pile of warm metal shavings. Discuss." That was the entire exam. Another famous examiner, on the faculty at MIT for many years, would set exams of this kind:

> There are fifteen important concepts in this course. Discuss any thirteen of them, outlining key ideas and providing proofs as time permits.

Max Zorn at Indiana University went one better. He wrote an exam in a graduate course that said only

$$\sin z$$

As a graduate student at Harvard, Ed Dunne determined to go to a math conference in California. Indeed, he was invited to present a paper at the conference. He was an impoverished student, and he certainly didn't have the resources to buy a ticket. He humbly made an appointment with the math department chairman, Heisuke Hironaka (1931–), Fields Medalist, and asked the great man if departmental support were available.

Hironaka asked him whether attending this conference was really important to him. Dunne said, "Yes," and Hironaka took out his personal checkbook and wrote him a check for the plane fare.

Statistician John H. Curtiss (1909–1977) said that

> You will recall that it has been said that World War I was a chemist's war and that World War II was a physicist's war. There are those who say that the next World War, if one should occur, will be a mathematician's war.

When he was nearing retirement from Harvard, George Mackey was extolling to a group of young mathematicians the virtues of the school. It had the best faculty, the most stimulating curriculum, and surely the most exciting students. Mackey was in a rapture over the many wonderful students—at all levels—whom he had had. One of the young listeners asked who were the standouts. Mackey said that, in the past forty years, there had been two undergraduates who were just extraordinary—clearly much better than anyone else. The first that Mackey mentioned was David Mumford, who was evidently a prodigy even as an undergraduate, and who was at that time both an endowed chair professor at Harvard and the chairman of the math department. "And who was the other?" queried the young fellow. "Oh, he is a plumber in Chicago."[1]

It is the custom at Harvard for there to be two "thesis defenses" for each Ph.D. The first is an informal gathering of the faculty (*not* the candidate), at which the thesis advisor explains to some of his colleagues what the thesis is about. This is, in part, to guarantee that the actual defense (the one at which the student *defends* his thesis) will go smoothly. At one such preliminary defense, the colleagues gave the thesis advisor (David Mumford) a particularly hard time. At the end, one of the august faculty told Mumford, on the q.t., that this was not a very impressive thesis. Finally Mumford said, "OK, so it's not the best thesis I've ever written."

On another occasion, Mumford was one of two faculty who gave an oral exam to some poor student. The young fellow was terribly intimidated, and became increasingly nervous. By the end of the session, he could hardly string words together to form a sentence. But the professors passed him, shook his hand, and the ordeal was over. The other professor left Mumford's office, upon which Mumford pulled out a bottle of scotch and told the student to drink up.

G. H. Hardy (of Hardy/Littlewood fame) wrote an article in 1909, in the *Mathematical Gazette*, about what he perceived to be "school mathematics." These days we are inundated with articles about mathematics education. These cover topics ranging from group learning to self-discovery to

[1] Evidently the second student, brilliant at mathematics, came from a great Chicago plumbing family. He had planned all along, after his time at Harvard, to go into the family business.

education by computer. Hardy's subject, by contrast, was stated definitively in his title: "The Integral $\int_0^\infty \frac{\sin x}{x}\,dx$." In the piece, Hardy lovingly describes seven different methods for evaluating this famous integral. He gives each of them a number of marks, ranging from 28 to 108 (more marks is a worse score). He thereafter does a reprise, and re-assesses his evaluations. His summatory remarks are these:

> And this fairly represents my opinion of their respective merits: possibly, however, I have penalized Nos. 2 and 5 too little on the score of artificiality and 4 too much on the score of theoretical difficulty. I conclude, however, with some confidence that Mr. Berry's second proof is distinctly the best. But whether this be so or not, one thing at any rate should be clear from this discussion: whatever method be chosen, the evaluation of the integral (1) is a problem of very considerable difficulty. This integral (and all the other standard definite integrals) lie quite outside the legitimate range of school mathematics.

And thus concludes G. H. Hardy's foray into math education.

In Fall of 1980, when I was a visiting faculty member at Princeton University, I was asked to serve on a qualifying exam committee. As noted elsewhere in this book, this means that I and two other faculty members (in this case Bill Browder and Gerard Washnitzer) were locked in a room with the candidate for three hours, and we could ask him whatever we wanted. This particular student did not fit the usual stodgy Princeton prototype: he wore a dashiki, and had his hair in dreadlocks.

The magic moment came around, and Browder (who was both the chairman of the committee and the thesis advisor of the student) said, "OK, your two advanced topics are differential topology and Lie groups and you are also responsible for real analysis, complex analysis, and algebra. Where do you want to begin?" The student smiled and said, "Differential topology is my thing. Let's start with that." So Browder posed a standard question—one that he had treated many times in class. The students eyes became wide, his whole body stiffened like a board, and he fainted dead away. Of course we faculty were not prepared for such an eventuality, and we stared at each other in bewilderment. After a few moments the student shook himself, got up off the floor, and assured us that everything was fine. Let the mad revels proceed.

William Browder

So Browder asked him another basic question from differential topology. The student's eyes grew wide like saucers, his whole body stiffened, and he fainted dead away. This time, after a few moments, he said, "Let me lie here for a few minutes and pull myself together. I'll be fine." So we twiddled our collective thumbs for a while and let the student recover. Then we went for round three.

Browder asked the student yet another basic question from differential topology. Again the students eyes dilated, and he fainted. Browder sprang up, called a halt to the proceedings, and walked the student to the infirmary. The upshot, after a diagnosis, was that the student had not eaten in three days nor slept in two. Another oral exam was scheduled. This time we all gathered together and the student decided to start with analysis (my bailiwick). I asked him to tell me something about the zero set of a holomorphic function. He said, "I don't know what you mean." I said, "Well, is it open or closed or connected or what?" He said, "It's either empty or it's everything." I think it would have been better if he had fainted. But we plunged on, and things went from bad to worse.

The poor fellow quit the graduate program within a year.

In 1984 I had the privilege of joining a group of American mathematicians to visit China. We spent much of our time at Beijing University (or the University of Peking). The person who looked after our daily needs, kept us comfortable, and planned our touring was Chen Zhenya. Chen was a top-level administrator in the mathematics department. His background was an undergraduate degree in mathematics. His special gift was great good humor, a fanatical attention to detail, and a wonderful and generous personality. Chen and I became fast friends. At some point during our month-long visit, Chen made it clear to me that he desperately wanted to spend time in the United States. This desire was partly motivated by curiosity; but the primary interest was that the imprimatur of time spent in the U.S. would immensely enhance his prestige back home in China. Thus Chen would get a better position, a better apartment, and a better life for his family. I determined that I would find a way to help Chen.

There was, however, an obstacle to our plans. In those days there were certainly funds to arrange for Chinese scientists to visit the U.S. A great many of them wanted to do so, and a great many of them did. But there was no program, and certainly no funds, for university staff to visit the U.S. When I returned to my job at Penn State in the U.S., I puzzled over how to concoct a way to help my new friend Chen Zhenya. Finally I hit upon it. I had a meeting with the Head of Penn State's Restaurant and Hotel Management Program. Together we cooked up a curriculum in "Applied Mathematics and Hotel Management" and we sent an invitation, on official Penn State University stationery, to Chen Zhenya to be the flagship student in our new effort. Chen, being a clever guy, was able to parlay this pseudo-document into permission and funding to visit the U.S.

Altogether Chen Zhenya spent more than two years with us. At Penn State, Chen took courses in mathematics and hotel management. He allowed that he could not make much sense of what the hotel people were talking about, but he enjoyed the math courses. When I moved to Washington University in St. Louis he came along, and he served as our Colloquium Chairman for a time. He also was the Assistant to the Librarian. On his return to China, Chen Zhenya became the Head of the International Office at Beijing University. He has an entire building for his staff and he runs a huge empire that oversees all foreign visitors and conferences at the university. Another success story for mathematics.

Our visit to China lasted a month. We worked hard for three weeks, giving lectures and teaching. The fourth week was our reward: we got to tour China. One stop was in Xian. We got to see the hundreds of terra cotta soldiers that had recently been unearthed, and we got to eat at the Liberation Street Dumpling Restaurant. All most delightful. At the end of the day we were exhausted and looked forward to flopping in our hotel. But there was a problem. It was Chinese National Day, the first to be celebrated since the time of Mao Tse Tung (1893–1976), and there was a huge commotion. Communist officials had come from all over, and all the hotels (and restaurants too) were full. Of course the person who was serving as our tour guide was too polite to tell us any of this, but he was scurrying around trying to make things right. He was able to arrange for us to participate in a fancy state banquet, with many Communist officials and other dignitaries. We hardly fit in, being all dusty and dirty and not speaking any Chinese. But the food was wonderful and we had a splendid time. Then we were taken to a Chinese discothèque where people lamely attempted to dance to what could only be described as bowdlerized Lawrence Welk music. This was OK for a short while, but we were kept there until midnight, wondering what was going on. Finally our tour guide picked us up, apologized for the delay, and promised to take us to our accommodations. We wearily piled into the bus and then we were taken to what looked like a pretty decent hotel. A huge sense of relief flooded the group. We went to our rooms, showered up, and were sitting around watching TV and getting ready to hit the sack when we noticed that there were a non-trivial number of people standing around who looked like burly guards. At a certain point, they made it clear (with motions rather than words) that they wanted us to go to bed. So we did. The final straw was when we observed that the rooms locked on the *outside* and could only be unlocked from the outside. We were housed in an insane asylum.

We managed to survive this ordeal intact. The next morning we were taken to a street market. In China in 1984 this was quite fascinating; in those days, one did not see such free enterprise in Peking. One had to travel some distance from the capital in order to detect this sort of capitalistic initiative. It was fun to see all the different vegetables and wares. Way at the end of the market was an emaciated, shirtless old man loudly wailing some chant and waving a rattle or noise-maker in the air. I went to investigate and was presented with this scene: He had a row of 25 rat tails laid out in front of him. He had a small pet rat running in circles on top of a circular platform. On a simple scale, with weights, he was weighing out careful portions from a pile of pink crystals. For each portion, he would tear out a page from an old

Chinese edition of Euclid, would wrap it up, and would sell it. All quite bizarre. I endeavored to take a photo, but the man became so upset that I had to cease. He did not even want me to observe what he was doing, but I persisted. And finally I figured out what was going on. The man was selling strychnine (rat poison). The 25 rat tails represented some of his triumphs. The little rat running around was his mascot. And the copy of Euclid was a source of free paper.

André Weil's sister was the noted mystic Simone Weil (1909–1943). She died quite young, while André died in his old age in 1998. It was natural that the *New York Times* would interview André about his sister. In fact it was an unending source of irritation to the great mathematician that that was *all* they ever wanted to ask him about. On one occasion, they asked him what he thought she was, or what she represented. He replied, "I don't know what she was, but whatever it was, she was most certainly a saint."

Tsuneo Tamagawa (1925–) is Professor of Mathematics at Yale University, the namesake of Tamagawa numbers. He likes to classify mathematical conversations into three categories:

- A *class one* conversation is one about theorems, proofs, conjectures, and mathematical methods.
- A *class two* conversation is one about mathematicians, mathematical gossip, or mathematical politics.
- A *class three* conversation is one about TeX or graphics packages or printers.

Along the same lines, my colleague Richard Rochberg (1943–) is fond of observing that, twenty-five years ago, if you heard a good math talk, then you might go up to the speaker afterward, request a preprint, look it over, and say, "Lemma 3 is particularly interesting. How do you prove it?" Today, if you hear a good math talk, then you might go up to the speaker afterward, request a preprint, look it over, and say, "How did you format Lemma 3? Which TeX font did you use for your running heads? How did you import your PostScript graphics?"

My colleague Nik Weaver (1969–) was a young Assistant Professor teaching calculus. He determined to teach his students the rigorous definition of continuity—using ε's and δ's—and he told them they would be quizzed on it on Friday. He would ask them to write down the correct definition and they were to do it. He pounded this fact into their heads. Friday rolled around, Nik gave the quiz, and then he later sat down to grade it. About halfway down the stack he came across a paper that provided this definition:

> For every $\varepsilon > 0$ there is a $\delta > 0$ such that you can draw the graph without lifting your pencil from the paper.

One year at MIT they decided that calculus would henceforth be taught non-rigorously. The dreaded ε's and δ's would no longer be part of the course, and that was departmental policy. Naturally some professors took umbrage with this new edict. One professor walked into his calculus class on the first day and announced, "In this class we will make use of certain constants called σ and τ. We define a limit as follows: For every $\tau > 0$ there is a $\sigma > 0$ such that… .

One of the big ideas in the subject of Finance in the past few decades is the Black/Scholes theory of option pricing. Nowadays a graduate-level measure theory course is liable to have as many economics, business, and finance students as mathematics students—just because the Black/Scholes theory relies on stochastic integrals.

Fisher Black (1938–) and Myron Scholes (1941–) were awarded the Nobel Prize for their work. This event was announced at a general faculty meeting at Stanford University, which was Scholes's affiliation at the time. Subsequently, a big article lionizing the event was written for one of the campus newspapers. One of the editors used a spell-checker rather ineptly, and every occurrence of "Myron Scholes" in the published article was replaced by "moron schools."

Persi Diaconis (1945–) is an original and pioneering statistician, holder of the MacArthur Prize. In fact when he won the Prize and was asked to comment on it he said, "If they give you a piece of paper to hang on the wall, that

Persi Diaconis

is of course very nice. But if they also give you $300,000 then people really pay attention." I once asked Persi about his name. He said, "I'm called 'Diaconis' because my father was Greek. I'm called 'Persi' because my mother was crazy." Persi has broad and diverse interests, including magic. In fact he ran away from home at the age of fourteen to follow a professional magician around the country. During that time he had sporadic and erratic experiences with various colleges. At some point he decided he wanted to go to graduate school in the mathematical sciences. But he did not know anyone; he had no idea whom to ask for a letter of recommendation. Persi turned to his friend Martin Gardner, who was well known at the time as the math writer for *Scientific American* and who is also a magician. Gardner said he would try to help, and that he happened to know a faculty member in statistics at Harvard (also, as it happens, a fellow magician). Gardner wrote to that august gentleman that he didn't know much about mathematics, but that this young fellow Diaconis had created three of the best ten magic tricks of the previous decade. Persi was admitted, and the rest is history.

Solomon Lefschetz was famous for heckling colloquium and seminar speakers. Ralph Boas got so tired of hearing this taunting that he resolved to heckle Lefschetz. Attending one of Lefschetz's lectures, he waited until Lefschetz mentioned a "Hausdorff space" and asked,"What is a Hausdorff space?" (of course Boas already knew the answer perfectly well). For the rest of the year, Lefschetz referred to Boas as "the man who doesn't know what a Hausdorff space is."

It is said that, late in his life, Hilbert was reading a paper and got stuck at one point. He went to his colleague in the office next door and queried, "What is a Hilbert space?"

Wiener was seriously concerned about the impact of scientific work—particularly his own—on society. He was, after all, the inventor of the subject and the term "cybernetics." He refused to address a symposium at the Harvard University Computation Laboratory on January 8, 1948, claiming that the devices under discussion were "for war purposes." In a letter published in the January, 1947 issue of the *Atlantic Monthly*, Wiener stated, "I do not expect to publish any future work of mine which may do damage in the hands of irresponsible militarists." He went on to say that, "The practical use of guided missiles can only be to kill foreign civilians indiscriminately, and it furnishes no protection whatever to civilians in this country."

When I was an undergraduate, a fellow student and I decided to learn what a Riemann surface was. We spent an afternoon going from door to door in the math department at the University of California at Santa Cruz, asking each professor in turn to explain the concept. Everyone seemed to know what one was, but nobody seemed to be able to explain it. Finally we screwed up our courage to ask the distinguished visitor of the week, Andrew Ogg (1943–) of U. C. Berkeley. After all, it was his lectures that had spurred our quest in the first place. So we went to his office and put the question to him. He spent several minutes thinking about the matter. We became hopeful, because we expected that he was endeavoring to formulate

Steven G. Krantz (as an undergraduate at UC Santa Cruz)

his answer in just the right way. Finally he picked up the chalk, made a careful dot on the blackboard, and said, “It’s obvious.” Then he left the room.

The computer science program at Carnegie-Mellon University is certainly one of the very best in the country. Graduate students in the Department of Computer Science claim that the only way they can survive their Ph.D. studies is with “Zac ‘n Jack.” This is a carefully measured mixture of Prozac and Jack Daniels.

A noted Japanese mathematician was teaching a calculus course at a large Midwestern university. He lectured by filling the several blackboards at the front of the room with long rows of tiny but neatly written script—uttering hardly a word. One day half an hour went by, and there were a great many dense paragraphs recorded on the board. Suddenly, the lecturer looked up with surprise and embarrassment on his face. He turned to the class, bowed, and said, “I’m sorry, I’m sorry” repeatedly and ran back and forth across the front of the room erasing little words and replacing them with other little words. This took some time. After about five minutes the lecturer turned to the class and offered his abject apologies yet again. “I am so sorry. I forgot that I was in the subjunctive mood.”

One of the special privileges of academic life is to direct Ph.D. theses. It is a way to stay alive mathematically, to nurture the next generation of mathematicians, and to create a school of thought. Some Ph.D. theses are more of a trial to direct than others. Hilbert said that supervising a Ph.D. thesis is just like writing a paper under exceedingly trying circumstances.

It takes a certain amount of courage for a graduate student to approach a senior mathematician and ask him to direct his/her Ph.D. thesis. The following story is told about Sammy Eilenberg (1913–1998) at Columbia University. A student came to him and said, "Sir, I'd like to write a Ph.D. thesis." Said Sammy, "Why don't you?"

Marshall Stone (1903–1989) of the University of Chicago is reputed to have said to students wanting to write a thesis with him, "If I had any good problems, I'd work on them myself."

In 1961, Paul J. Cohen used a daring new method called "forcing" to prove the independence of the continuum hypothesis from the other axioms of set theory. During the period that he was working on the problem, he was in the habit of telling people that he was working on the Riemann hypothesis. After he got the solution to the independence problem, he was invited to the Collège de France to give a big public lecture on his result. Of course all the logicians in France were present. Cohen began his lecture by saying, "This is a problem which has been around for some time—over thirty years, in fact—and nobody has really made much progress on it. That's not very surprising, though, because since Gödel there hasn't been anyone first rate working in this area."

Once Cohen had his independence proof nailed down, he took his manuscript to Princeton for verification by Gödel, the master. By that time in his life, Gödel had become quite neurotic, indeed phobic. Cohen went to Gödel's house, but the old man would not open the door. Instead, a hoary hand slipped out the door to grab the manuscript. Several days later, Gödel pronounced the result correct, and Cohen was invited in for tea.

There are several legends about how Paul Cohen came to work on the independence of the continuum hypothesis. After all, he was trained as a harmonic analyst under Antoni Zygmund, and his early published work is on multiple Fourier series, spectral synthesis, and other analytical prob-

Paul Cohen

lems. Those who knew Cohen in graduate school attest that he had a long-standing interest in the foundations of mathematics. One story is that Cohen was sitting one day in the coffee room with a number of other mathematicians. He asked what problem he should work on that would garner him the most immediate fame and glory. Various candidates were considered, including Hilbert problems and the Riemann hypothesis. It was finally decided that the independence of the continuum hypothesis was the ticket, and eighteen months later he had solved it.

Harish-Chandra (1923–1983) was a permanent member of the Institute for Advanced Study. Although he wrote many papers, it is often claimed that he wrote just one; for each was a direct continuation of the last. Paper n would have no introduction; it would just pick up where paper $n-1$ had concluded (and, by the way, paper $n-1$ had no particular conclusion; it just stopped).

Harish-Chandra is credited with developing the theory of harmonic analysis on reductive p-adic groups. The Plancherel formula in the p-adic context is one of his triumphs. Harish-Chandra was fond of the unifying principle that whatever is true for a real reductive group is also true for a p-adic reductive group. He called this the "Lefschetz principle," and it led him to his proof of the Plancherel formula.

At the end of Harish-Chandra's 1972 lecture series at a conference in Williamstown, Massachusetts, he told this story that he attributed to Claude Chevalley (1909–1984). The tale relates to the time before Genesis when God and his faithful servant, the Devil, were preparing to create the universe. God gave the Devil a free hand in building things, but told him to stay away from certain objects which He, Himself, would attend to personally. Chevalley's story was that semisimple groups were among those special items. Harish-Chandra added that he hoped that the Lefschetz principle was also on that proprietary list.

In the 1960's, André Weil attended a lecture by Mitchell Taibleson at the Institute for Advanced Study. He kept interrupting Taibleson, insisting that Taibleson (an analyst) use a certain notation to express the modular relationship. Taibleson said that he preferred to emphasize the analysis rather than the algebra. After several interruptions, Weil said, "This is my notation, and I'm giving you permission to use it."

On another occasion, Weil was preparing a lecture in which he had to use some Fourier analysis. One particular point kept confusing him, and he consulted Taibleson several times for help. In fact he went to Taibleson repeatedly on the same point. He eventually declared that he had it straight. But, during the lecture, Weil once again became confused. Looking at the audience in exasperation, he finally declared, "If you don't understand this, consult your analyst."

When a member of the public (a layman) is asked to name a famous mathematician, the most commonly heard answer is "Albert Einstein." Those of us who know better realize that Einstein was a physicist, and even he protested that his mathematical abilities were modest. Weil found the whole situation irritating. He was fond of saying that Einstein's only contribution to mathematics was to invent the Einstein summation notation.

It is said that in the early part of the twentieth century Stefan Bergman took a class from Erhard Schmidt (1876–1959) about L^2 of the unit interval. Remember that, in those days, the concept of Hilbert space had not yet been invented or developed. So it was natural for a mathematician to study a particular Hilbert space. In any event, Bergman's command of German was as muddled as his command of several other languages. Legend has it that Bergman misunderstood what Schmidt was talking about, came away with the notion that he was discussing square integrable holomorphic functions on the unit disc, and thus the Bergman space (the space of square-integrable holomorphic functions on a domain) was born.

In any event, the mid-1970's were a heyday for the Bergman kernel, one of Bergman's chief inventions. Thanks to work of J. J. Kohn, Norberto Kerzman (1943–), and Charles Fefferman (1949–), there was much activity in the subject, and a 1975 conference on several complex variables had many talks on ideas related to things that Bergman had studied or developed. Bergman attended the conference, and was pleased to see his efforts finally being recognized for their full worth. His wife Edy was with him (usually seen chasing him across the room while shouting, "Stevie, Stevie!"), and she too appreciated the recognition. I sat next to Bergman at many of the principal lectures. In each of these, he listened carefully for the phrase, "and in 1923 Stefan Bergman invented the kernel function." Bergman would then dutifully record this fact in his notes—and nothing more. I must have seen him do this twenty times during the three-week conference.

There was a rather poignant moment at the gathering. In the middle of one of the many lectures on biholomorphic mappings, Bergman stood up and said, "I think you people should be looking at representative coordinates" (also one of Bergman's inventions). Most of us did not know what he was talking about, and we ignored him. He repeated the comment a few more times, with the same reaction. Five years later, S. Bell (1954–), E. Ligocka (1947–), and Sidney Webster (1945–) found astonishing simplifications and extensions of the known results about holomorphic mappings using—guess what?—Bergman representative coordinates.

Salomon Bochner (1899–1982) was a mathematician at Princeton University. Along with Lefschetz, he is generally credited with building the Princeton mathematics department into the world center that it is today. The custom at Princeton is that, when a professor retires, he relinquishes his

Salomon Bochner

office. Princeton is just too busy a place to provide a camping ground for its retired faculty. So Bochner arranged a new home for himself: he moved to Rice University in Houston. Indeed, Bochner is credited with contributing significantly to the build-up of the Rice mathematics department. One day a young John Polking, still on the faculty at Rice today, went up to Bochner at tea and asked what he had done in his Ph.D. thesis. Without hesitation, Bochner said, "I invented the Bergman kernel." A bold statement indeed. But one can check the *Mathematische Annalen* paper of Bochner from 1915 and indeed see a vestigial form of the Bergman kernel.

When Bochner went to Rice, after his retirement from Princeton, he became chairman of the department. He ran the show with an iron fist, and this style extended to the way that he offered jobs. A mathematician of my acquaintance (who has since left math and become an economist) applied to

Rice for a job. He subsequently received a phone call—directly from Bochner—offering him a position. Bochner told him in no uncertain terms that he was to accept or decline the offer *on the spot*. He was to be given no time for consultation or consideration. The candidate said, "Rice is my first choice. But I am entertaining several other options, and I would like to see what they have to say before I accept your kind offer." Bochner became irritable, and demanded to know from whence these other offers emanated. The candidate told him, upon which Bochner said, "Anyone who would consider offers from those places doesn't deserve a job at Rice." And he hung up the phone.

The story has a happy ending. A few years later the same candidate applied to Rice again, was given an offer, and he accepted. He spent several happy years at Rice University.

Another young mathematician whom I know got the same offer from Bochner, and he was so insulted that *he hung up on Bochner!*

There is a German fable about Barbarossa (the Emperor Frederick I). He died while on a crusade, and is buried in a far-off land. Legend has it that he is still alive, and sleeping in a cavern of the Kyffhäuser mountain. But he will awaken, even after hundreds of years, and return to the Fatherland when the German people need him. Somebody allegedly asked Hilbert, "If you would revive, like Barbarossa, after five hundred years, what would you do?" Hilbert replied, "I would ask, 'Has somebody proved the Riemann hypothesis?'"

George Green (1793–1841) grew up in Sneinton, England, and was the son of the man who built the tallest, most powerful, and most modern windmill (for grinding grain) in all of Nottinghamshire. As a child, George Green had only 15 months of formal education; he left the Goodacre Academy at the age of 10 to help his father in the adjoining bakery.

After the publication of his first scientific paper in 1828, when George Green was 35 years old, Green became an undergraduate at Gonville and Caius College, Cambridge. After obtaining his degree, Green became a Fellow of the college and continued his scientific research in mathematics and physics. Unfortunately, his health failed and he returned to Sneinton, where he died in 1841. At the time of his death, George Green's contributions to science were little known nor recognized. Today we remember George Green for Green's theorem and the Green's function of partial differential equations.

Today—after many mishaps, fires, and disasters—Green's Mill has been restored to full working order. One can visit the Mill and the Visitors' Center, learn of the history of the mill and of George Green, and purchase a variety of souvenirs. These include a sample of flour produced at the mill, a key fob, and a pencil emblazoned with Green's theorem.

One of the great theorems of twentieth century mathematics is the Nash embedding theorem: that any Riemannian manifold can be embedded isometrically into Euclidean space. John Nash published this result in the *Annals of Mathematics* in 1958. Unfortunately, this event coincided almost exactly with the beginning period of his mental breakdown, a condition that incapacitated him for more than 30 years. [Many readers will know that Nash has now recovered; he functions normally, and is thinking about mathematics. A few years ago he was awarded the Nobel Prize in Economics. Today there is a first-run movie about John Nash called *A Beautiful Mind*, starring Russell Crowe.] As a result, the write-up of his theorem that Nash submitted to the *Annals* was a chaotic mess—certainly unpublishable in that form. The referee of the paper was Herbert Federer (1920–), and Federer voluntarily rewrote the entire paper. The form in which the paper now appears is due largely to Federer.

The short story "The Devil and Simon Flagg" (1954) by Arthur Porges tells of a mathematician who has devoted his life to endeavoring to prove Fermat's Last Theorem (FLT). He is obsessed with the problem, to the point that he loses his friends and loses his wife and (very nearly) loses his sanity. At one point he is approached by the Devil. The Devil offers to help him solve the problem if only he will relinquish his immortal soul for all eternity. The mathematician readily acquiesces. So the Devil says, "Fine. I can always use another soul. Now, what do I need to know to solve this little problem?" The mathematician replies that first the Devil needs to learn calculus, and gives him a text. One day later the Devil returns and says that this was trivial; he read the book and now he has mastered calculus. What next? The mathematician gives him a book on number theory. Two days later the Devil is back, dusting off his hands and saying that that was simple enough. What next? Then he is given a book on complex analysis. That, too, is the work of just a couple of days. To make a long story short, after about ten days the Devil is ready to tackle FLT. He and the mathematician begin to collaborate furiously. Several months go by, and in the closing scene of the story it is late at night. The Devil and the mathematician are sitting at a table laboring away by candlelight and the Devil is heard to say, "I just need this one lemma..."

G. H. Hardy said that, in his youth, he was once walking through a thick fog with a clergyman and they saw a boy with a string and a stick. Hardy's clergyman compared this scene to the invisible presence of God, which can be felt but not seen. "You see, you cannot see the kite flying, but you feel the pull on the string." Hardy knew, however, that in a fog there is no wind so kites cannot fly.

Stan Ulam was once at a party in Princeton. He spied a wizened old man, gaily drinking champagne and with a young girl on his knee. He asked his host von Neumann who that was. "Oh! don't you know? He is von Kármán [student of Hilbert], the famous aerodynamicist." He added, "Don't you know that he invented consulting?"

At the height of his fame, Theodore von Kármán (1881–1963) was a professor both at Aachen in Germany and at Cal Tech in Pasadena, California. Since he consulted for the airlines, he usually flew for free. So he was constantly shuttling back and forth between his two venues. He frequently gave the same lecture in both places. One day he was lecturing at Cal Tech and the faces in the audience seemed a bit more puzzled than usual. He suddenly realized that he was lecturing in German. He apologized profusely and told the audience that they should have stopped him long before. The students were painfully silent, but finally one young man spoke up: "Don't get upset, Professor. You may speak German, you may speak English, we will understand just as much."

Norbert Wiener is reputed to have always sat in lectures in a semi-somnolent state except when he heard his name (at which he would suddenly

Stan Ulam

jump up, then sit back in a very comical way). Wiener was quite ambitious about his place in the history of mathematics. Once Saunders MacLane was lecturing on something algebraic, and Wiener was sleeping as usual. At the end, MacLane said very loudly, "And thus we see that this subject has absolutely nothing to do with ERGODIC THEORY." Wiener woke up and began to talk about ergodic theory.

Norbert required constant reassurance about his creative ability. A few weeks after their first acquaintance, Wiener asked Stan Ulam, "Ulam! Do you think I am through in mathematics?"

Ulam once spotted Wiener in front of a bookstore with his face glued to the window. When Wiener saw Ulam, he said, "Oh! Ulam! Look! There is my book!" Then he added, "Ulam, the work we two have done in probability theory has not been noticed much before, but see! Now it is in the center of everything."

In the 1950's, Norbert Wiener and Norman Levinson and some of the younger mathematicians at MIT (one of whom told me this story) were in the habit of playing bridge at lunch. Each time he laid down a card, Norbert would say, "Did I do the right thing? Was that the best possible play? Am I a good bridge player?" And, each time, Norman Levinson would patiently assure Wiener that nobody could have done any better.

Alfred North Whitehead was a noted mathematician and philosopher, co-author with Bertrand Russell of *Principia Mathematica*. He was once asked by an acolyte which was more important, ideas or things. "Why, I would say ideas about things," was the immediate reply.

One of the peculiarities for which Paul Erdős was most famous was his use of language. For him, an *epsilon* was a child, a *boss* was a wife, a *slave* was a husband, *noise* was music, *preaching* was lecturing, *the supreme fascist* was God, *capture* was marriage, *to die* was to stop doing mathematics, *to leave* was to die, *to be liberated* was to get divorced, and so forth. Once Ulam walked with Erdős on the beach, and Erdős stopped to caress a child. "Look, Stan! What a nice epsilon," cried Erdős. Ulam's taste was for the child's mother, an extremely beautiful young woman who was sitting near-by. He said, "But look at the capital epsilon." This made Erdős blush with embarrassment.

One evening Erdős was at a party, enjoying the company and the many small epsilons. At one point he got a very sad look on his face and stood leaning against a pillar, looking quite bereft. A friend asked him what was wrong and he said, "One of my theorems just died."

Erdős used to joke about his mortality. He autographed a book of mine with a long acronym about his being "almost dead," "half dead," and so forth. He used to like to say that he was 2.5 billion years old, because when he was a child scientists said the earth was 2 billion years old but more recently they have said it is 4.5 billion years old. Erdős was fond of giving a talk entitled "My first 2.5 billion years in mathematics."

One day Erdős noticed that the audiences at his talks had been getting larger and larger. He claimed that the audiences were filling halls so large that his old and feeble voice could not be heard. He speculated as to the cause. "I think," he said, "it must be that everyone wants to be able to say 'I remember Erdős; why, I even attended his last lecture!'"

Erdős had the habit of phoning mathematicians all over the world (on somebody else's nickel, of course) without any regard for time zones, or when people might be asleep or otherwise occupied. Of course his intent was to discuss mathematics. He knew by heart the phone number of every mathematician of his acquaintance. [One story tells of Erdős looking up ten numbers in the phone book, being distracted for 30 minutes by a conversation, and at the end still knowing the numbers cold.] But he did not know anyone's first name. The only person whom he called by his Christian name was Tom Trotter; Erdős called him "Bill."

On one occasion, Erdős met up with a mathematician and asked him where he was from. The reply was "Vancouver." "Oh," said Erdős, "then you must know my good friend Elliott Mendelson (1931–)." After a moment's silence the reply was, "I *am* your good friend Elliott Mendelson."

My friends tell me of the time Erdős took a taxi from Indiana University to Purdue University (a distance of 100 miles), found he had no money in his pockets, and threw himself on the mercy of the math department. The chairman had to take up a collection to pay the taxi driver.

There are many Paul Erdős stories that illustrate his eccentricity and self-absorption. This one is particularly quaint. During World War II, Erdős was walking on the Long Island beach with Shizuo Kakutani (a Japanese) and Arthur Stone (1916–2000, an Englishman). They strayed into a restricted area and as a result were apprehended. The authorities separated the three mathematicians and interrogated them. Erdős's interview went something like this:

Q: What were you doing on the beach?
E: Walking.
Q: Where were you going?

E: Nowhere.
Q: Then what were you doing on the beach?
E: We were discussing mathematics.
Q: What about mathematics?
E: We were considering a conjecture.
Q: What conjecture?
E: It doesn't matter. It was false.

At one point John von Neumann was endeavoring to recruit Stanislaw Ulam to work on the atomic bomb project at Los Alamos. At the time, von Neumann was not authorized to say exactly what the project was or where the work would take place. He simply told Ulam that it was interesting and exciting work, and important for the war effort. It would involve mathematics, physics, and engineering. Ulam said, "Well, as you know, Johnny, I don't know much about engineering or experimental physics, in fact I don't even know how the toilet flusher works, except that it is a sort of autocatalytic effect." At this von Neumann winced, and his expression became quizzical. A little later Ulam said, "Recently I have been looking at some work on branching processes." John von Neumann looked at Ulam suspiciously and with a wan smile. It turned out that the word "autocatalytic" had been used in various theories connected with creating the A-bomb, and branching processes were a device used to study neutron multiplication. Poor von Neumann evidently feared that Ulam had been reading his mail.

John von Neumann was the brains and inspiration behind the Princeton computer—one of the first stored-program computers—known simply as the IAS computer (this was in an era when the great computing machines were known as MANIAC, ILLIAC, and the like). It is said that he knew all the electronic parts and he supervised their assembly. When the machine was near completion, he made fun of it at his own expense. He said, "I don't know how really useful this will be. But at any rate it will be possible to get a lot of credit in Tibet by coding "*Om Mane Padme Hum*" (the Tibetan chant "Oh, thou flower of lotus") a hundred million times an hour. It will far exceed anything a prayer wheel can do."

G. H. Hardy and J. E. Littlewood had perhaps the most famous and certainly the most prolific collaboration in mathematical history. It lasted 35 years, and produced 100 papers and a book; they created many important techniques and ideas in analysis and number theory that are still used today.

Hardy and Littlewood had four axioms for their collaboration. Before enunciating these, we should note that most of the mathematical communications between Hardy and Littlewood were by handwritten letter—today known as *snail mail.*

> **Axiom 1:** When one wrote to the other, it was completely indifferent whether what he wrote was right or wrong. As Hardy put it, otherwise they could not write completely as they pleased, but would have to feel a certain responsibility thereby.
>
> **Axiom 2:** When one received a letter from the other, he was under no obligation whatsoever to read it, let alone to answer it. They followed this rule strictly. Béla Bollobás observed Hardy getting thick letters from Littlewood daily. He simply threw them into the corner, declaring that he may want to read them some day.
>
> **Axiom 3:** Although it didn't really matter whether they both thought about the same detail, it was preferable if they did not.
>
> **Axiom 4:** It was a matter of complete indifference if either one of them made no identifiable contribution to a paper that was to appear under their joint authorship. That way there would never be any priority disputes.

Littlewood gave a good deal of thought to what attributes of life were most conducive to creative work. In his senior years, he always took Sunday off, never thought a moment about mathematics, and thoroughly relaxed. Then, on Monday morning, he was always in the mood to plunge into work and claimed that he usually came up with a good idea. In his own words, "for serious work one does best with a background of familiar routine, and that in the intervals for relaxation one should *be* relaxed… for me the thing to avoid, for doing creative work, is above all Cambridge life, with the constant bright conversation of the clever, the wrong sort of mental stimulus, all the goods in the front window."

Littlewood was a witty guy, and considered to be charming company. As an example, he liked to say that, since childhood, he had imagined that the biblical character Samson had *pulled* the pillars, and was puzzled how he could reach. He had also seen a picture confirming this thought. But in the Hollywood film rendition Samson used the sound rock-climbing technique of *pushing* the pillars. Abram Besicovitch once asked Littlewood what God was doing before the Creation. He said, "Millions of words must have been written on this; but he was doing Pure Mathematics and thought it would be a pleasant change to do some Applied."

Harry Williams is the one who told Littlewood about Limbo, a benevolent invention of St. Thomas Aquinas. As Littlewood put it, "You have all the natural pleasures, but may not twang the harp. Good enough for me."

A friend once asked what Hardy was like. Littlewood laughed. He asserts that he *should* have said, "All individuals are unique, but some are uniquer than others."

J. E. Littlewood

Littlewood gave considerable thought to how how the mind came up with ideas and, more generally, how creativity worked. He said that, "Incubation is the work of the subconscious during the waiting time, which may be several years. Illumination, which can happen in a fraction of a second, is the emergence of the creative idea into the conscious. This almost always occurs when the mind is in a state of relaxation, and engaged lightly with ordinary matters." For example, he noted that Hermann Helmholtz (1821–1894) usually got his ideas while walking during periods of rest. Littlewood noted particularly that the activity of shaving could be a fruitful source of minor ideas. He used to postpone it, when he could, until after a period of work. He summed things up with, "Illumination implies some mysterious rapport between the subconscious and the conscious, otherwise emergence could not happen. What rings the bell at the right moment?"

On another occasion, Littlewood noted that, "The higher mental activities are pretty tough and resilient, but it is a devastating experience if the drive does stop, and a long holiday is the only hope. Some people do lose it in their forties, and can only stop. In England they are a source of Vice-Chancellors."

Certainly Littlewood railed strongly against the common vices. He admonished that there should be no smoking until the day's work is complete. There were good arguments against regular smoking: you are merely normal when smoking and miserable when not. Littlewood had been in the habit of smoking 16 pipes and 4 cigars per day. A friend of his had to give up tobacco for 4 or 5 weeks following upon a severe bout of flu. The friend found that he then completed a difficult paper in one third the time. Hearing this, Littlewood engaged in the struggle for abstinence. He won, and found thereafter that his work was considerably more efficient and effective.

Speaking of creativity, Littlewood once noted that there are the "great" creations, of something totally new and unexpected, and also of considerable importance and seminal.[2] We should all feel, he went on, that the difference is one of kind and not of degree. He then posed the queston: Is the difference

[2] Physicist Richard Feynman (1918–1988) also considered this notion. He said that there were certain scientists who simply seemed to work much harder than the rest of us, and we could aspire to be like them. Others were "magicians," and one had no notion of where their ideas came from.

between a difference of degree and a difference of kind a difference of degree or a difference of kind? The *answer*, of course, is elementary.

Gian-Carlo Rota (1932–1999) was one of the many contributors to the monumental work *Linear Operators* by Nelson Dunford (1906–1986) and Jacob Schwartz (1930–). This is a three-volume, 3500-page treatise that has exerted enormous influence over the subject of functional analysis. One of Rota's tasks was to check the exercises in Chapter 3. He labored long and hard over the job, and was embarrassed that he could not do Exercise 20 of Section 9. He finally told Dunford of his trouble. The two of them, and several other members of the group working on the book, spent a few hours trying various tricks; but they could not do the problem either. Next day they consulted Schwartz, and he also could not do the problem. A few years later, Dunford was giving his graduate course on linear operators. He assigned Exercise 20, and one of the first-year graduate students not only solved it but constructed an elegant theory around it. He was Robert Langlands (1936–).

A bit later Rota was called in to an audience with Dunford and Schwartz. They told him that his next task was to do the problems in Sections G and H of Chapter 13. Rota noticed with chagrin that they were not giving him the exercises in Section I, on the use of special functions in eigenfunction expansions—a particular favorite of his. In fact they were giving that section to an undergraduate whom they described with the phrase "You will never find a better undergraduate in math coming out of Yale." In fact it was John Thompson (1932–), who later went on to study with Saunders MacLane and to win the Fields Medal.

Incidentally, Dunford hated semi-colons, and forbade that any appear in the book. Whenever he encountered a chance semi-colon his face would get red. The graduate students working on the project were scared to death that Dunford would catch them trying to slip in a semi-colon.

When I was a graduate student at Princeton I once overheard D. C. Spencer telling an admiring group, "I just love sheaves. They have algebra this way (and he sliced his hand up and down) and topology this way (and he sliced his hand from left to right)." At the time I did not know what a sheaf was, so could not appreciate the wisdom of Spencer's remarks. Now I know that he was right.

D. C. Spencer

Spencer was a native of Colorado, attended college there, and returned there upon his retirement. He was admitted to Harvard Medical School (following in the footsteps of his physician father), but walked out of the first lecture he ever attended and went to MIT, where he fell under the spell of Norbert Wiener. After receiving a B.S. in aeronautical engineering at MIT, he went to Cambridge University in England to study mathematics (more on this elsewhere). The bulk of his professional career was spent at Princeton University. Spencer professed to never being comfortable in the rather formal and stuffy atmosphere of that university town. He said that he always felt like he was walking around with his fly open.

One of Spencer's many achievements was his long and successful collaboration with Fields Medalist Kunihiko Kodaira (1915–1997). They wrote twelve papers that laid the foundations for the theory of deformation of complex structures in several complex variables. These ideas are still studied intensely today. Kodaira left the Princeton mathematics department in 1963, in part because Princeton would not match an offer from another university. Spencer was so angry that he resigned his position and moved to Stanford. He was persuaded to return to Princeton five years later.

Spencer served as mentor and teacher to many distinguished mathematicians, including J. J. Kohn, P. A. Griffiths, and John Nash. He "didn't think [Nash] was going to get anywhere." After Nash was stricken by mental illness, Spencer made great efforts to see that Nash received first-class treatment.

D. C. Spencer died on December 23, 2001.

When D. C. Spencer first went to Cambridge to study mathematics (over his conservative father's objections), his intention was to study classical, British-style applied mathematics. One day he was walking across the quad and saw a youngish fellow, in a three-piece suit with his shirt sleeves rolled up, scaling the side of a building. Spencer thought this was great stuff. He threw down his books and immediately ascended in hot pursuit. The two fellows met at the top, shook hands, and went off for a pint. It turned out that the first climber was John Edensor Littlewood. Spencer ended up writing his thesis with Littlewood.

In 1957, Gian-Carlo Rota used to lunch with Oskar Zariski (1899–1986), who liked to use the opportunity to practice his Italian. One day they were sitting in the Harvard Faculty Club and Zariski announced in a loud voice, "Remember! Whatever happens in mathematics happens in algebraic geometry first."

The Institute for Advanced Study was founded in Princeton, New Jersey in 1930. It opened its doors, so to speak, in 1933. The Institute was generously endowed with funds from Louis Bamberger of Bamberger's Department Stores (now a part of Macy's). The intellectual guiding light behind the concept of the Institute, and the man who put it all together, was Abraham Flexner. Flexner had been characterized as the "hanging judge" of American higher education. He was mainly famous for the "Flexner report," which was an exposé of quackery and fraud in the nation's medical schools. He was also an effective fundraiser for American universities that adhered to his strict standards.

The Institute is not formally a part of Princeton University, although it shared space on the Princeton campus for a few years before it became permanently situated a few miles down the road. In choosing the initial leading lights to be founding members of his institute, Flexner wrote to the top six or eight people in each of three or four disciplines, asking each of them to identify the top six or eight people in their discipline. The only field in which he got any real consensus was mathematics. That is why the Institute began with the School of Mathematics. The first permanent members of the Institute were Albert Einstein, James Alexander (1888–1971), Marston Morse, Oswald Veblen, John von Neumann, and Hermann Weyl. The permanent Institute grounds are also a game preserve; they are quite lovely. There are apartments for the visitors—those staying at the Institute for anywhere from a few weeks to one or two years. An interesting feature of these apartments is that there are only single beds, even though many are occupied by married couples and families. The theory is that either there was a terrific fire sale at which the Institute got all the beds, or else the powers that be want the visiting members to concentrate on their work.

It is no exaggeration to say that Albert Einstein was a legend in his time—long before he went to the Institute for Advanced Study. After all, he was the person who taught us that light rays could be bent by the sun's gravity, that space was curved, and that the universe has four dimensions. When he went to visit at the house of J. B. S. Haldane (1892–1964) to stay the night, Haldane's daughter took one look at the man and promptly fainted dead away.

Einstein was once asked where he kept his laboratory. The great man smiled, took a fountain pen out of his breast pocket, and said "Here."

Paul Erdős was a "natural mathematician." It was never how much he knew; it was rather how clever he was and how hard he tried. My colleague Robert H. McDowell (1927–) tells of working with Erdős and a group of mathematicians at Purdue. They gathered in the morning, and someone went to the blackboard to present a problem: "Let X be a Hausdorff space…" "What is a 'Hausdorff space?'" asked Erdős. "Now assume that X is separable…" "What is 'separable?'" asked Erdős. And on and on. But by the end of the day they had solved the problem and they could write a paper. The next day, in high spirits from their earlier success, they gathered again. Someone went to the blackboard to present a problem: "Let X be a Hausdorff space…" "What is a Hausdorff space?" asked Erdős. And so it went.

In a similar vein, in 1939 at the Institute for Advanced Study, Hurewicz raised the question of what is the dimension of the rational points in Hilbert space. Erdős was not quite sure what Hilbert space was, and he had never heard of "dimension." These ideas were explained to him, and he soon solved the problem. He published the result in the *Annals of Mathematics*.

In 1948 Paul Erdős was in residence at the Institute for Advanced Study and of course Atle Selberg (1917–) was a permanent member. Selberg got a promising idea for obtaining an elementary proof of the prime number theorem (that is, a proof that does not use complex analysis or other tools outside of elementary number theory). Paul Erdős was able to supply an important step that Selberg needed. Later, Selberg was able to modify his proof so that it did not require Erdős's idea. Unfortunately, a terrible priority dispute erupted between Erdős and Selberg. One version of the tale is that Selberg was visiting at another university, sitting in somebody's office and having a chat. Another mathematician walked in and said, "I just got a postcard saying that Erdős and some Norwegian guy that I never heard of have found an elementary proof of the prime number theorem." This really set off Selberg, and he was then determined to write up the result all by himself.

Dorian Goldfeld (1947–) has taken pains to interview all survivors who participated in or witnessed the feud between Erdős and Selberg. It is clear that nobody was wrong and nobody was right. Both Erdős and Selberg contributed to this important discovery, but they had a significant clash of egos and of styles. Irving Kaplansky (1917–) was in residence at the Institute in those days and witnessed the feud in some detail. He tells me that at one point he went to Erdős and said, "Paul, you always say that mathematics is part of the public trust. Nobody owns the theorems. They are out there for all to learn and to develop. So why do you continue this feud with Selberg? Why don't you just let it go?" Erdős's reply was, "Ah, but this is the prime number theorem."

In the popular conception, a mathematician toils away at his desk, perhaps awash in papers and books, slide rule in hand and computer crunching away. Those of us in the profession know that—at least for many of us practicing mathematicians—nothing could be further from the truth. Often a solitary walk in the woods is the best way to tackle a problem. Perhaps hanging out in the coffee room and jawing with colleagues is the way to do it. For others, a swim in the ocean or a session with the nautilus machine will set the brain cells afire. For Steve Smale (1930–), basking on the beach at Copacabana was just the ticket. There he pulled off one of the real coups of modern mathematics: he created cobordism theory and proved the Poincaré conjecture in all dimensions 5 and greater.

R. L. Moore was a point-set topologist at the University of Texas. He had many excellent graduate students, and is particularly remembered for the so-called Moore method of teaching.

An unpublished biography of R.L. Moore tells of Moore's rough-and-ready childhood schooling in a one-room schoolhouse in late nineteenth century East Texas. One Fall, the headmaster was overheard on the first day of the school year saying to a parent "Do you want me to teach the earth round or flat?—I can do it either way."

R. L. Moore was a mainstay of the University of Texas math department from the 1920's to the 1970's. Possessed of a strong and single-minded personality, he developed a special method of teaching that has been successfully practiced and disseminated by his students and followers.

The Moore method worked like this—and it was practiced in all his classes, from the most elementary undergraduate to the most advanced graduate. Students would show up on the first day, usually with no notion of what Moore was like or what he expected. He would hand out a single mimeographed sheet with some definitions and some axioms and some theorems stated on it. And he would say, "Who would like to go to the board and prove the first theorem?" Of course Moore was usually greeted by bewildered silence. Many of his students did not know the meaning of the words "definition" and "axiom" and "theorem" and "proof," nor did they know the basic rules of logic. But Moore could take it. He sat and he waited. Often the first hour would go by without a word spoken. And so would the second. And many times the third as well. Eventually, some brave soul would go to the blackboard and attempt a "proof" of Theorem 1. Moore would rip that person apart. And that set the tone for the class.

In each session of R. L. Moore's class, some student would go to the board with a proof he had concocted, and everyone else in the class—student and professor alike—would examine it critically. There were no holds

R. L. Moore

barred. Moore would not *assist* the students in creating a proof, but he would *respond* to whatever they said.

The good thing about Moore's classes is that his students came to *possess* the mathematics. If you put it incorrectly on the blackboard, then you were subjected to merciless criticism. If you got it right, then you received quiet but sincere approbation.

Moore did not allow his students to read books or papers (this included even his graduate students). Indeed, Moore saw to it that the mathematics library got no funds for new books. As legend had it, if he found a student of his in the library he would throw that person (bodily) out the door. Students were to create the mathematics themselves—from whole cloth. The class worked best if the students began with approximately equal abilities. They were not allowed to read, and they were not allowed to collaborate outside of class. Moore was merciless in weeding out those students who did not cooperate or did not fit.

Mary Ellen Rudin, one of Moore's most prominent and successful students, does not recommend the Moore method. She says, in part, "All you get [from this method] is a (probably overconfident) uneducated ignoramus." By contrast Paul Halmos, an eloquent avatar of good teaching, says, "The Moore method is, I am convinced, the right way to teach anything and everything—it produces students who can understand and use what they have learned." Moore's teaching method—and his techniques for selecting the students whom he wanted in *his classes*, and whom he would mentor—caused emotions to run high in the University of Texas Mathematics Department. It is said that "... the department was literally divided into two camps: the Moore and the anti-Moore. The Moores had the fourth floor, and it was armed: guns in the drawers up there."

When R. L. Moore was 67 or 68 years old, he had a student who was quite proud of his athletic and physical accomplishments. He was always crowing about them. Moore challenged the student: he did one-armed push-ups, with his feet up on a chair, so that he was with his head angled down during the display. At the end, he jumped up and had not even broken a sweat or lost wind. The student was duly impressed. It has been said of Moore (long before the notion of "empowerment" became trendy) that one of his gifts was that he gave students faith in their own (mental) strength.

R. L. Moore is remembered as being strong-minded about everything. When he planned to teach a class, he saw to it that all relevant books were taken off the library shelves. For many years, he was one of the most influential members of the University of Texas Mathematics Department. He convinced the administration that they needed no money in the budget for the library. There was no value to purchasing mathematics books.

According to the unpublished biography of Moore, he was eventually forced to retire. The assertion is that, when the University of Texas became a "mega-university" with 30,000 or more students, then it could not tolerate the creative teaching style of a man like R. L. Moore. Too many constraints were put on him, and he had to step down. Fortunately, the Mathematical Association of America has made a delightful film about Moore, and anyone today can witness his unique teaching style.

Paul Halmos offers sage counsel to those who want to consult, or to give advice to those in the applied world. Number one: your main function is as a psychotherapist. Just listen to the description of their problems, nod your head a lot, ask some simple-minded questions. This will help them to formulate their problem more precisely. It is really therapeutic. They will be immensely grateful. If you can actually answer their questions, it is sometimes (but not always) useful for you to do so. But, says Halmos, never, ever give a counterexample. They will neither appreciate it nor understand it. I in fact had this experience as a graduate student. A physics student asked me whether one could factor a polynomial of two variables into linear factors. "No," I said. "Look at $x^2 + y$. He got quite angry and said, "I'll just have to speak to someone who knows what he is talking about." I should have followed Paul's advice and helped the guy reformulate his question. Instead I said, "To hell with him."

Richard Feynman tells, in his biography, *Surely You're Joking, Mr. Feynman,* of being called in on a consulting job. He was presented with some incredibly complicated blueprints for a nuclear reactor that he simply could not understand. He stared at the huge printout in complete bewilderment for several minutes, sure that he was going to make a fool of himself. Then he pointed to something and said, "Is that a window?" The engineers in the room then got quite excited and said, "Oh, we see what you mean. The ventilation should go here. Then the pressure can dissipate there. If we do that, then the fission…" Soon the engineers had their problem solved, and they assigned all the credit to Feynman's sage advice.

After he had been in science for many years, and had served both as collaborator and consultant on many occasions, Feynman offered this advice: "If anyone asks me a question, I always say, 'Differentiate under the integral sign.' More than half the time this will solve their problem. And, even if it doesn't, they will think you are a really smart guy."

Alfred Jules Ayer (1910–1989) was one of the leading lights of the logical positivist movement in Britain in the early part of the twentieth century. Bertrand Russell and other luminaries of mathematics and logic were adherents. Ayer's book *Language, Truth, and Logic* was a blockbuster which had the philosophy community up in arms. In it, Ayer argued cogently that all ethical statements were just "emoting," that there was no way to verify what was in the world around us, and no way to establish any commonality among people. What we saw and what we thought and what we sensed were, for all we knew, distinct; and there would never be any way to prove otherwise.

One day one of Ayer's students came to the distinguished man's office in a state of some agitation. He said, "Professor Ayer, is it true that you think that man has no commonality?" "Yes," answered Ayer. "And is it also true, Professor Ayer, that you believe that there is no way to confirm or deny that our sensual experience coincides with the product of our thoughts?" "Most certainly," answered Ayer. "And finally, sir, do you believe that any ethical statement is nothing more than mere emoting?" "Indeed I do," replied Ayer. "Good," said the student. "Because I am running off with your wife."

W. H. Young (1863–1942)—father of L. C. Young (1905–2000), who is mentioned elsewhere in this book—was a professor at U. C. Berkeley. A distinguished mathematical analyst, Young developed an interest late in his life with theoretical issues of typesetting. It is said that when he taught his analysis course he would devote the bulk of the time to teaching his students a language he had developed for expressing all of mathematics in ASCII code that could be written left-to-right across the page. What most people do not realize is that, more than sixty years ago, W. H. Young anticipated TEX.

It was arranged, around the time of the start of the Cold War, for Bertrand Russell to give a talk to a conservative women's club in Britain. The topic was "Politics in England." This was not a marriage made in heaven, for Russell was known to be a left-wing philosopher and an outspoken atheist. He endeavored to deliver his message in a calm and deliberate manner, but the ladies were not at all receptive. Indeed, at the end of the presentation they jumped up and attacked Russell with their parasols and other loose goods lying around. The Sergeant-at-Arms came to the rescue. On the one hand, he wanted to save Russell; on the other hand he wanted to be gracious with the ladies. He shouted, "Be gentle! This man is a great mathematician." No good. They continued to pummel Russell. Then the Sergeant-at-Arms cried, "Be gentle! He is a great philosopher!" To no avail. They continued to assault their guest. Finally the Sergeant-at-Arms shouted, "Be gentle! His brother is an Earl." That did the trick.

Bertrand Russell was once giving a public lecture in Britain on the subject of astronomy. He of course tendered the Copernican theory of how the earth orbits around the sun. He went on to assert that the sun, in turn, orbits around the center of a vast collection of stars which is called our galaxy. At the end of the lecture there was a question-and-answer session. A little old lady in the back of the lecture hall got up and said, "What you have told us is rubbish. The world is really a flat plate supported on the back of a giant tortoise." Russell could not resist giving a supercilious smile as he replied, "But what is the tortoise standing on?" The old lady shot back, "You're very clever, young man, very clever. But it's turtles all the way down!"

Miriam H. Young, writing in *The Arithmetic Teacher*, notes the following progression of ideas:

Plato: "God ever geometrizes."
Jacobi: "God ever arithmetizes."
Dedekind: "Man ever arithmetizes."
Cantor: "The essence of mathematics is in its freedom."

Great Failures

William (Willy) Feller (1906–1970) and his wife were once trying to move a large circular table from their living room into the dining room. They pushed and pulled and rotated and maneuvered, but try as they might they could not get the table through the door. It seemed to be inextricably stuck. Frustrated and tired, Feller sat down with a pencil and paper and devised a mathematical model of the situation. After several minutes he was able to prove that what they were trying to do was impossible. While Willy was engaged in these machinations, his wife had continued struggling with the table, and she managed to get it into the dining room.

Before the days of NSF grants, Casper Goffman used to make extra money in the summer working in a shoe store. Later he was lucky enough to get a job consulting for an aerospace firm. One summer they were testing the prototype of a new aircraft. One of the standard tests is to move the wings up and down a short distance, imitating the motion of the wings in flight, tens of thousands of times to test the plane for stability and durability. The engineers constructed a machine to effect the motion of the wing, a second machine to record (on moving graph paper) the position of the wing against time, and a third machine to record (on moving graph paper) the velocity of the wing against time. After running a test, the engineers brought the two printouts to Goffman for comment. Of course both were periodic waves, and the engineers observed that they seemed to be the *same wave*, except that one was translated $\pi/2$ units to the left of the other. Could Goffman provide an explanation? He said, "Well, the derivative of sine is cosine." The engineers looked unenlightened and said, "What does that have to do with it?"

Abram Besicovitch, in spite of his apparent powers, was modest. On his thirty-sixth birthday, he convinced himself that his best and most intensive years of research were over. He said to some friends, "I have four-fifths of my life." Twenty-three years later, when in 1950 he was awarded the Rouse Ball Chair of Mathematics at Cambridge, someone reminded him of his frivolous remark, pointing out that he had written more than half of his papers in the intervening time—many of them among his best. Besicovitch's reply was, "Well, numerator was correct."

There is a story about Sir Michael Atiyah (1929–) and Graeme Segal (1941–) giving an oral exam to a student at Cambridge. Evidently the poor student was a nervous wreck, and it got to a point where he could hardly answer any question at all.

At one point, Atiyah (endeavoring to be kind) asked the student to give an example of a compact set. The student said "the real line." Trying to play along, Segal said, "In what topology?"

Sir Michael Atiyah

Kakutani always liked to blame Andrew Gleason (1921–) for the Vietnam War. The story is that, in 1940, when Gleason was a freshman at Harvard, his roommate was McGeorge Bundy (1919–). Bundy's intention at that time was to study mathematics. But he found Gleason to be so brilliant, and so intimidating, that he quit math and instead studied politics.

As we all know, McGeorge Bundy went on to become one of the architects of the Vietnam War. And so Kakutani's accusation arose.

The following excerpt from J. L. Doob's (1910–) review of the book, *Norbert Wiener 1894–1964,* by P.R. Masani (1919–1999) is quite telling:

> Norbert Wiener: a precocious genius whose father once shook him by the shoulders and told him if he ever amounted to anything, the credit belonged to the father; who fostered a naive *enfant terrible* image, for example ostentatiously reading popular magazines while sitting in the front row of joint Harvard/Massachusetts Institute of Technology seminars; who was so unsure of himself that his MIT colleague and former student Levinson had as one of his duties to assure him frequently that he was a fine mathematician; so naive in his need of appreciation that he had the habit of asking other mathematicians what he was, a mathematician, an engineer, or a statistician—the desired answer was of course, 'all three'; who on the other hand was self-assured enough to write to Hitchcock, the master director of suspense movies, offering to send him a movie plot; who felt himself so far out of the science establishment that he resigned from the U. S. National Academy of Sciences; so isolated that he stalked the halls of American Mathematical Society meetings peering nearsightedly to find a friendly face; so estranged from humanity that he wrote a novel whose principal character was an unorthodox genius who finally killed himself in despair at his lack of appreciation by his colleagues (themselves readily identifiable living mathematicians, thinly disguised by pseudonyms, and so unhappy at their book characterizations that they influenced publishers to reject the novel); whose feeling for physics and appreciation of Lebesgue integration was so deep that he was the first to understand the necessity and the proper context for a rigorous defini-

tion of Brownian motion, which he then devised, going on to initiate the fundamentally important theory of stochastic integrals; who, however, was so unfamiliar with the standard probability techniques even at elementary levels that his methods were so clumsily indirect that some of his own doctoral students did not realize that his Brownian motion process had independent increments; who was the first to offer a general definition of potential theoretic capacity; who, however published his probabilistic and potential theoretic triumphs in a little-known journal, with the result that this work remained unknown until too late to have its deserved influence.

Halmos once attended a lecture in Russia that he characterized as "awful"—incoherent, disorganized, and impossible to understand. Another member of the audience confessed that he had just turned off his hearing aid during the talk. He also said that a friend of his had tried to get him to cut down on his drinking, claiming that alcohol consumption affected the hearing. After some thought, the man replied, "On the whole, I think I prefer what I drink to what I hear."

Paul Halmos was John von Neumann's assistant at the Institute for Advanced Study. One day they were working on a problem together and, using some fiendishly clever new ideas, they solved it. Of course they were delighted and von Neumann told Halmos to copy down the solution; he, von Neumann, had to go catch a train. Halmos said not to worry; the proof was emblazoned on his frontal lobes. It was late Friday and he would go drink some beers. He would write down the proof the next day. Come morning, Halmos could not reproduce the proof. And the janitor had erased the blackboard. There was nothing left. Halmos spent a miserable weekend trying to reconstruct the argument that he and von Neumann had concocted, but to no avail. When von Neumann returned a few days later, Halmos sheepishly told him that he had "lost" the proof. Unfortunately, von Neumann had forgotten all about the problem, and he, too, had forgotten the proof. He and Halmos had to solve the problem all over again from scratch.

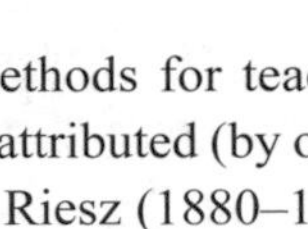

There is much discussion these days of different methods for teaching mathematics. A style that is rarely seen today has been attributed (by one of his students) to the famous Hungarian analyst Frigyes Riesz (1880–1956). Riesz would march into the lecture hall, followed by an Associate Professor and an Assistant Professor. The Associate Professor would read to the class from Riesz's famous book and the Assistant Professor would copy the words and symbols onto the blackboard. Professor Riesz would stand front and center, hands behind his back, and nod sagely.

Arguably the heyday of analysis at the University of Chicago was the 1950's. Calderón and Zygmund were developing their theory of singular

Frigyes Riesz

integrals, there were many brilliant students, and things were booming. Zygmund was writing his *magnum opus*, *Trigonometric Series*; this book still stands today as the paradigm for what every hard analyst must know. It is said that, in those days, if a student went to Zygmund and asked to write a thesis, the canonical response was, "OK. First read my book *Trigonometric Series*, then come back and talk to me." This task took about two years, and filtered out the faint of heart. When Harold Widom, now a distinguished analyst, got this reply from Zygmund, he said, "I have a wife and a child. I don't have time to screw around." In particular, he did not have two years to spend reading Zygmund's book. So he went and wrote a Ph.D. in algebra. When he completed that exercise, he immediately returned to his beloved analysis, never to deviate since.

One day it was announced at Stanford that Marcel Riesz (Frigyes's brother) would give a series of four lectures. The first day was warm, the room was large and crowded, and Riesz soon removed his jacket and was lecturing in his shirtsleeves and suspenders. A bowl of water and a sponge had been provided for cleaning the blackboard. After he had filled up the board, Riesz motioned imperiously to Gabor Szegő (1895–1985), who sprang up and washed the blackboard while Riesz stood by and observed the proceedings. It should be borne in mind that, at this time at Stanford, Szegő was a very distinguished and polished faculty member who wore tailored suits; he was regarded with awe by students and faculty alike. He was also Executive Head[1] of the department. It was startling for one and all to see Szegő now in the role of a European-style assistant to Riesz.

George Pólya was in the audience, and he had brought Felix Bloch (1905–1983) with him to hear a distinguished fellow Hungarian speak. He was quite embarrassed by the whole performance, and he kept muttering *sotto voce* apologies throughout.

The Princeton University campus is a very traditional and elegant place. Some of the buildings are more than 200 years old, and the atmosphere is one of serene dignity. One of the more charming features of the campus is

[1] Usually the "Head" or "Chairman" of a department is appointed by the Dean with advice of the department's faculty. The Executive Head was empowered directly by the President and Board of Trustees.

Marcel Riesz

that there is a small railroad station with a little train that runs regularly on a spur out to the main train line between New York and Philadelphia. This little electric train is called "the Dinky." The Dinky has been running for many, many years, and those wishing to enjoy a night in New York or Philly use it as a convenient means of travel. That portion of the campus that adjoins the train station and the Dinky is quite "old school." There are eating clubs (the Princeton alternative to fraternities) and other austere buildings.

One day, about twenty-five years ago, it was proposed that a convenience store be built next door to the Dinky station. It was to be a branch of the Wawa chain of markets. A great hue and cry was raised when the tentative plans were unveiled. How would such a tacky edifice fit into the Princeton environment? This sounded like the sort of thing one might do at a public institution—perhaps that *state school* 20 miles up the road. This certainly was *not* for Princeton. But the forces of business won out over the forces of tradition, and the Wawa was built. It became the most popular late-night hangout in Princeton, for it was open 24 hours and was a great place to get a sandwich, drink a beer, play a video game, or just be cool.

One night some Princeton students were making the scene at the Wawa, and they got a bit drunk. One of the inebriates climbed on top of the parked Dinky and, through an unfortunate mishap, he was electrocuted. Now this was a very high voltage line that he was touching. He did not die, but he lost all four of his limbs.

Very sad, and a number of law suits ensued. The hapless student sued the train company, the Wawa (for selling him the beer), the university, and ... he sued his physics professor. What? The student claimed that his professor had given him a "gentleman's *B*" in freshman physics, thereby misleading the student into thinking he understood electricity. This misdirection no doubt led to the student's unfortunate mishap. The university settled out of court.

Certainly one of the most remarkable mathematical places in the world is the Mittag-Leffler Institute in Djürsholm, Sweden. The institute is in what was actually Gösta Mittag-Leffler's house. The Mittag-Lefflers were wealthy, in large part because of his wife's family money. It is, like its neighbors, a very large house, in the grand style of the wealthy families of the nineteenth century. In fact most of the other houses nearby are today occupied by ambassadors to Sweden from other nations.

Scholars have not had time nor opportunity to study Mittag-Leffler's artifacts, so most of them still sit on the shelves of Mittag-Leffler's house, much as they did when he was alive. His study has vertical shelf files with his correspondence from Karl Weierstrass (who was Mittag-Leffler's teacher), Jacques Hadamard (1865–1963), and other contemporary mathematicians.

A large room off the foyer, possibly a former library, has many enclosed shelves at floor level. On these shelves are boxes and boxes of family photographs of Mittag-Leffler and his friends, family, and associates.

One of Gösta Mittag-Leffler's most famous associates was Sonja Kowalewska (1850–1891). She lived in Mittag-Leffler's house for some time, and it was rumored that they were intimate (she was *not* his wife). Among Mittag-Leffler's photographs are snapshots taken at a Halloween party. In one of these, one can see the lovely Sonja Kowalewska dressed up as a kitty kat.

Sonja Kowalewska

Alfred Nobel (1833–1896) was the inventor of dynamite. He built a huge factory to manufacture the product, and became exceedingly wealthy by marketing his invention. As a legacy to his accomplishments and his vision, he established the Nobel Prize. There is a story that Alfred Nobel did not specify in his will for a Nobel Prize in mathematics in large part because Mittag-Leffler ran off with his wife. It is true that Mittag-Leffler was a prominent public figure—the most visible scientist in Sweden at the time—and something of a dandy and a lady's man. But Nobel never had a wife, and there is no evidence that he ever even had a girlfriend. What seems more probable is that Nobel was a practical man of the world and was (just as is the public today) rather unaware of mathematics. It never occurred to him to establish a Nobel Prize in mathematics. Other explanations have also been offered. Mittag-Leffler was a worldly, well-educated man-about-town who was frequently in the newspapers—in fact Mittag-Leffler's scrapbooks are on prominent display in the lobby of the Mittag-Leffler Institute, and each occurrence of his name in an article or clipping is underlined in red. By contrast, Alfred Nobel was a modest, down-to-earth businessman. As a result, it seems that Nobel bitterly resented Gösta Mittag-Leffler; it is rumored that he did not set up a Nobel Prize in mathematics because Mittag-Leffler might have won it.

Gösta Mittag-Leffler was quite upset that there was no Nobel Prize in mathematics. As a result, he established his own prize, called the Mittag-Leffler Prize. The recipient would receive a medal twice as large as the Nobel medal, a complete leather-bound edition of *Acta Mathematica* (the mathematics journal that Mittag-Leffler founded), and a fancy dinner cooked by a celebrated chef. One of the first recipients was Charles de la Vallée Poussin (1866–1962), who was vacationing in the Swiss Alps at the time. Mittag-Leffler arranged for the journals, the medal, and the meal all to be carted up to de la Vallée Poussin in the Alps.

Unfortunately, the Mittag-Leffler Prize was only awarded twice before it went bust. It seems that Mittag-Leffler had endowed the prize by investing in German World War I bonds and the Italian railroad system.

While there is no Nobel Prize in mathematics, other prizes have been created that are part of the Nobel trust. One of the most recent of these is the Crafoord Prize in Mathematics (and other sciences too). Created with funds from a wealthy Swedish pharmaceutical family, this prize has become a highly regarded and much-coveted encomium. It is awarded by the Swedish Academy (as is the Nobel Prize). The first mathematical Crafoord Prize was awarded to Louis Nirenberg (1925–) of the Courant Institute. On the day of the award, Nirenberg left Courant after a long day at work and was stopped by a *New York Times* reporter who asked how he felt about being the first recipient of the Crafoord award. Louis's reply was, "Is this some kind of joke?"

André Bloch (1893-1948) was a brilliant mathematician after whom Bloch's constant is named. His famous result is that any normalized, hyperbolic Riemann surface contains a univalent disc of radius at least $\mathcal{B}$, where $\mathcal{B}$ is some positive constant of indeterminate size. It is still an open problem to calculate the exact value of $\mathcal{B}$.

The interesting and tragic fact about Bloch is that, in 1917, he apparently slaughtered several members of his family with a cavalry sword. He was then incarcerated in an institution for the criminally insane for the remainder of his life. The details of the matter are murky, and there are many conflicting reports (see the article of Douglas Campbell in *The Mathematical Intelligencer* 7, issue 4, (1985), 36–38). One version of the story is that Bloch was on convalescent leave from his service in World War I. He became enraged that his brother (also a mathematician) was going to go to the Polytechnique while he (André) had to return to the grim trench warfare. So Bloch slaughtered his brother, his uncle, and his aunt. Other versions of the story assert that the point at issue was one of religion. Another is that he slaughtered his landlady because the noise in the building interfered with his mathematics. He seems to have escaped execution only because he was a war hero.

Be that as it may, Bloch conducted his entire professional life from his place of confinement (which may have been the Asylum at Charenton, where the Marquis de Sade had been held, or may have been at the "house of health" in Saint Maurice). He wrote a paper with George Pólya and was close to Jacques Hadamard and to Sjolem Mandelbrojt (1899–1983). Bloch published under several aliases, including René Binaud and Marcel Segond. He is said to have been a charming correspondent, and he received many invitations to dinner. His standard reply was "circumstances beyond my control prevent me from accepting your kind invitation."

The Ph.D. is a fairly recent invention; it is an outgrowth of the German intellectual modus of the nineteenth century. The tradition of having a formal thesis advisor is an even more modern invention. There are many living mathematicians—especially in Europe—who are virtually self-trained. Thus many famous mathematicians of times past had no Ph.D. One day someone approached one of these august gentlemen and asked, "How is it that you have no Ph.D.?" The gentleman drew himself up and said, "Who would examine me?"

The Ph.D. did not become established in Great Britain until the 1920's. Thus, for instance, G. H. Hardy did not have a Ph.D. His comment on the degree was, "A German invention, suitable for second-rate mathematicians and foreigners."

Hans Rademacher

Part of the ceremony of earning a Ph.D. is the "thesis defense." The Swedes have a particularly charming tradition for conducting this activity. Unlike our habit in the United States, in Sweden the candidate himself does not actively participate. Instead three Professors are enlisted. One is usually a distinguished foreign professor who is to read the thesis carefully and present the results in a public forum. A second professor, usually a Swede, is assigned to ask difficult questions of the first professor. A third professor, always a Swede, is assigned to make jokes and ridicule the entire proceeding. These days, the tradition continues, but (more is the pity) the role of the third professor has been eliminated.

In the post-World War II period, Paul Erdős was visiting Hans Rademacher at the University of Pennsylvania. When Erdős was getting ready to leave,

Rademacher indicated that he thought he had a proof of the Riemann hypothesis (RH). Erdős listened with interest to Rademacher's idea, but concluded by warning him that he thought it was flawed. He had better be careful. In those days communications were more primitive than today: there was no Internet and certainly no e-mail. Even long-distance phone calls were a luxury in which most folks did not indulge. Overwhelmed by his excitement at proving RH, Rademacher sent out postcards to a number of his friends announcing his great breakthrough. Word spread of the achievement, and a small article even appeared in *Time* magazine. In the end, Erdős turned out to be right. The proof was flawed. Rademacher had to withdraw his claim.

The University of Washington in Seattle has a lovely physical layout. The view of Lake Washington is spectacular, and can be enjoyed from the faculty eating club and from many of the offices. Salaries at this university are generally rather low, and my friends there say that they pay $20,000 per year for the view.

Similarly, people at the University of Wisconsin used to say that the beautiful physical location, on the shores of Lake Mendota, constitutes part of the faculty salary. Stanislaw Ulam used to say that every time he looked at the lake it cost him $2.

Stefan Banach (1892–1945) penned the important book *Opérations Linéaires* in which he develops the idea of a complete normed linear space. He calls these "spaces of type (B)," perhaps because he hoped that they would someday be named after him. And, indeed, they are today known as Banach spaces. In one of his books, Norbert Wiener called the spaces "Banach–Wiener spaces." This never caught on.

John von Neumann told marvelous stories of his travels. Once, at the request of friends, he bought several pounds of caviar in Moscow to bring back to the U.S. He asked a steward on the train to store the caviar in the icebox of the restaurant car. In the morning when they woke up, in Poland, they discovered that the restaurant car had been uncoupled at the Polish-Russian border. He returned to the States caviarless.

John von Neumann

One day Eugene Wigner (Nobel Laureate) and John von Neumann, as young men in school, went to a cafe in Budapest where billiards were played. They asked an expert waiter there if he would give them lessons at the game. The waiter said, "Are you interested in your studies? Are you interested in girls? If you really want to learn billiards, you will have to give up both." Wigner and von Neumann had a short consultation, and they decided they could give up one or the other but not both. They did not learn to play billiards.

In 1946 Stan Ulam became quite ill, and was out of commission for several months. At one point he told von Neumann of his illness: "I was given up for dead, and thought myself that I was already dead, except for a set of measure zero." John von Neumann laughed and asked, "What measure?"

In his younger days, Littlewood used to say that incompetent research students (i.e., Ph.D. students) reminded him of a general in *War and Peace*.

This general, disturbed by the groans of a soldier, calls to him, "My good man, do try to die more quietly!" When he heard that a mathematician had left Trinity College in Cambridge to be the Master of another College, he exclaimed, "What a come down!" Likewise, when he heard that the pianist I. Paderewski (1860–1940) had become President of the Polish Republic, he exclaimed, "What a come down!"

Littlewood once challenged Hardy to find a misprint in one of their joint papers. His collaborator was unable to do so. In point of fact, the misprint was in his own name: "G, H. Hardy."

G. H. Hardy, J. E. Littlewood, and some others were having a discussion of writing and style. Hardy said, "What's wrong with my style is not lack of expressiveness or lucidity, but vulgarity."

Hardy took great (sensual) pleasure in calligraphy, and it was he who always wrote the final copy of the joint Hardy/Littlewood papers in his elegant hand. He took pride in buying the most expensive paper for the work. If the first half-sentence of a page had an error, he would discard the sheet and start again.

Hardy felt strongly that the taste for old brandy was a pose, and he decried it as typical of a dishonest pretension to highbrow taste. When he encountered a similar posing, he would label it as an "old brandy stunt" or "o.b.s." For example, Samuel Butler's statement that Handel is better than Beethoven is an instance of o.b.s.

When a student, Hardy was invited by his tutor (Arthur Berry of King's College) to lunch. He thought hard about his letter of acceptance. He could not write "shall come with pleasure" because in fact he *had* to go. Similar logical contradictions caused him to reject all other formulations of the letter, and in the end he sent none. He just showed up for the luncheon. When he was told, "it is usual to answer an invitation" he could again think of no sincere answer, and he was too inarticulate to explain. He told this story, twenty years later, to Littlewood. At that time he would have been in command of the situation, but in the interval Berry had died.

Bertrand Russell is famous for co-authoring, with A. N. Whitehead, the book *Principia Mathematica.* Russell did the lion's share of the writing, for at the time Whitehead was a hard-working lecturer. Russell claimed that the task took so much out of him that he had never been quite the same again.

When Littlewood read Arthur Eddington's (1882–1944) *Report on Relativity* he was mightily impressed, and felt that it was the greatest intellectual advance and illumination that had ever happened. He explained it to Bertrand Russell, who at that time knew no physics. Russell was similarly staggered. He suddenly burst out with, "To think I have spent my life on absolute muck."

Littlewood relates that when Hermann Weyl taught in Zürich, his class of 100 or more dwindled to 1 student—Mrs. Weyl. He later got his teaching act together and became a most impressive instructor.

When I was a graduate student at Princeton, my teacher Frederick J. Almgren (1933–1997) came to me one day and asked if I knew a reference for a classical example of a pathological set in the plane. He knew that I had been spending a lot of time studying the many strange examples generated by Giuseppe Peano (1858–1932), Besicovitch, and others in the early part of the twentieth century. Of course I was flattered that one of my professors had sought my counsel. It turned out that, later that afternoon, L. C. Young was flying in from Wisconsin with a student who was to have his Ph.D. oral. Almgren was his second reader. Young knew, the student knew, and Almgren knew that Chapter 2 of the thesis was incorrect. The reference I had supplied gave the counterexample. Yet the student insisted on going through the charade of the final oral. Almgren encouraged me to attend the exam.

I did attend, and the student fought like crazy to avoid talking about Chapter 2. But the examiners insisted on it, the error was revealed, and the student was awarded a conciliatory Masters Degree.

Around the time of the "Chapter 2" debacle, the Princeton graduate students were going through an intense spate of practical joking (unfortunately begun by me). When Steven Weintraub (1951–) had his thesis defense, it was scheduled to be held in the Fine Hall tower, in the same room where the "Chapter 2" defense took place. In the middle of the exam, Weintraub turned to face his audience. Through the window he could see a placard suspended on strings from the floor above; it read "Mistake on page 72 of the thesis. You flunk." Weintraub burst into laughter. The examiners turned around to see the source of the levity, and then quickly grabbed for copies of the thesis to determine the error. The thesis had only 69 pages.

In those days, when I was writing my Ph.D. thesis, I came into work quite early each day. Nobody else was around. I could not help but notice that the Princeton fire marshalls were having a field day. Every week there were new fire doors, new signs posted giving occupancy limits, new fire extinguishers, and many other safety features added to enhance our lives. Feeling mischievous, I got a sign-making kit and made up a rather professional-looking sign that read

In case of fire remove all clothing. Don
fire-proof togas that will be provided.

And I hung the sign in the venerable Princeton tea room. It will tell you what sort of place Princeton is that the sign hung there for three years; nobody knew what to make of it.

In fact the first mathematician to take formal note of the "toga" sign was Guido Weiss, visiting from Washington University. Guido has played a seminal role in my life for many reasons, not least of which being that he attracted me to be a faculty member at Washington U. At tea in the Princeton Common Room, Guido turned to a group and asked very seriously, "Where is my toga?"

In those days, as a graduate student, I found Princeton most depressing. I could not understand most of what anyone was talking about, and I could never make heads or tails of the colloquia. But Guido Weiss had written a great book[2] —that I had recently read—and I was determined to go to his

---

[2] The book is *Introduction to Fourier Analysis on Euclidean Spaces*, joint with E. M. Stein. This work is a classic, and is still read extensively today. Elsewhere in this book is a story of the writing of that masterpiece.

talk and to learn something from it. I asked him beforehand what I needed to understand his talk. Without hesitation, Guido said, "A central nervous system." And he was right. His talk was lucid, comprehensible, and educational. He even went so far as to show respect and concern for my questions, and to give thoughtful answers. These are signs of a true scholar.

J. E. Littlewood tells of a former student who began brilliantly. He took a pure research post after getting his Ph.D. under Littlewood's direction. He then had six years of research. The work ultimately became dull, though there was much of it. When Littlewood met him again he was on the point of a nervous breakdown. The teacher then discovered that the young fellow had worked continuously for the six years for 365.25 days per year. Said Littlewood, "If he had done the work in reverse order he would have been a Fellow of the Royal Society."

Gian-Carlo Rota once witnessed a spirited discussion of artificial intelligence. The advocates of "hard AI" were painting a triumphal picture of the future of computer intelligence. In fact the interchange could hardly be called a discussion; the AI proponents were endeavoring to steamroll everyone else. Jacob Schwartz, Rota's teacher, was uncharacteristically silent. After a while, everyone turned to him. "Well," he said, "some of these developments may lie one hundred Nobel prizes away." People found this to be a calming statement, for it gave the AI people their due while still acknowledging the dreamy nature of many of their claims.

Rota went to graduate school at Yale. Traditionally, Yale has been especially strong in the humanities, with science playing second fiddle. The one really outstanding scientist at Yale in the old days was Josiah Willard Gibbs (1839–1903). In fact Gibbs is the man who invented the modern concept of vector. Gibbs served as professor at Yale without any salary. In those days (the late nineteenth century), teaching young men from the upper crust was not a salaried profession but a privilege for the happy few. One day Gibbs received an offer from Johns Hopkins University. It was an endowed professorship, presumably the one relinquished by J. J. Sylvester who had accepted a position at Oxford (after the religious vows for professors were dropped as a requirement). Gibbs was evidently pleased with the offer, and was inclined to accept it. One of his colleagues thought it would be disastrous if Gibbs left and he went to the Dean to formulate a plan to keep Gibbs at Yale. The Dean said, "Why, just tell him that you'd like him to remain at Yale." It turned out that this was just what Gibbs needed to hear. He turned down the Johns Hopkins offer and stayed at Yale for the remainder of his career.

One of the well-known endowed research Assistant Professorships in mathematics is the E. R. Hedrick Assistant Professorship at UCLA. Just who was Earle Raymond Hedrick? In the Math Library at UCLA is a file of his papers, and one can have a look. For one thing, Hedrick was the world's greatest expert on functions that satisfy the Cauchy-Riemann equations but are not holomorphic. *What?* Could this man be famous for papers that are incorrect? No, they are perfectly rigorous and true. But remember that, in Hedrick's day, there was no concept of "distribution derivative." What Hedrick found were examples of functions that satisfy the C-R equations in the pointwise sense; but they are so pathological that they are not distributions. In particular, Weyl's lemma does not apply and the functions in question are not holomorphic.

When Gian-Carlo Rota was a second-year graduate student, the secretary of Yale University was charged with finding an escort for Mrs. William Sloane Coffin, Sr., one of Yale's most generous benefactors. In the end, Rota was asked to do it. He escorted Mrs. Coffin on several evenings. Her black limousine would pull up in front of Woolsey Hall, where he would wait for her. She was old and feeble, almost blind, and he would help her carefully up the stairs and into the building. As they entered the rotunda, a high officer of the University would suddenly appear and rush over to greet her, brushing Rota aside. Rota's duties included walking Mrs. Coffin around during intermissions. Mrs. Coffin's two favorite topics of conversation were the bridges on the Arno River in Florence and the poetry of Ludovico Ariosto (1474–1533). Rota had to improvise, as he had never been to Florence. He did better with the poetry of Ariosto, as there was at that time no English translation of his poems.

James Alexander was a powerful and influential mathematician, but he was intensely shy and private. For example, he never took an assistant at the Institute. One year he announced a series of lectures on homology. He drew a large crowd for the first of these. He began at the beginning, very abstractly and axiomatically. He provided so many formal details that he did not get very far. At the next lecture, he announced that he was not satisfied with the axiomatic basis presented the first time around; so he would begin again. He gave a second formulation of his theory. Unfortunately, at his third lecture, the same thing happened. The fourth lecture was postponed—Alexander skipped a week. There were many more postponements and skipped weeks, but Alexander finally ended up giving six lectures, each starting over from scratch. Finally Alexander abandoned the course: he could not whip the subject into the shape that he wanted.

Paul Halmos is actually the subject of somebody's Masters thesis—not the *advisor*, you'll note, but the *subject*. This was a student of Speech, and her avowed goal was to eliminate Halmos's accent. She analyzed his speech patterns and set about systematically to eliminate all artifacts of the Hungarian argot. She taught him not to roll his "r"s. If you have ever conversed with Paul Halmos, you will know that she failed. Paul is still proud of his Hungarian burr.

Mathematicians are not always the most well-adjusted people. Sometimes they lose it, and tragedy can result. Here is the story of one murder in a mathematics department.

I first must supply a personal note. My Ph.D. thesis was based in part on work of Walter Koppelman (1929–1970) of the University of Pennsylvania. My source was a very brief research announcement that Koppelman had published in the *Bulletin of the AMS*. I could never find the promised subsequent paper that would fill in all the details, and I had to fill them in myself. I eventually went to my thesis advisor and asked him where the missing paper was. He said, "Oh, God. Don't you know?" And then he told me the sad story. There was a very unhappy graduate student at the University of Pennsylvania. He had had bad experiences with several thesis advisors (at least so he thought), the last being Koppelman. One day he went into the colloquium, shot the department chairman, shot Koppelman, and shot himself. Koppelman and the student died.

These days it is getting harder and harder to get NSF research grants. Money is tighter, there are more mathematicians, and there are more demands on the money—including mathematics institutes, educational projects, and other mathematical activities. One year recently a rather esteemed mathematician at MIT was turned down for his summer research grant. He became quite perturbed, as one can imagine, and he sued the NSF. This turned out to be a shrewd move, for all the relevant documents were confidential, and none could be released to any of the lawyers. In short, the suit could not be tried. As a result, our eminent friend was in the end awarded his grant.

When Erdős was a young man, still residing in Europe, he was an avatar of leftwing politics. He and his friends got together to discuss mathematics, but also to discuss socialism and how it was the only realistic solution to the problems of the world order. Of course being a leftist had its dangers, and Erdős was one of the first to spread the news when one of their number got nabbed by the police. "A. L. is studying the theorem of Jordan." is how Paul would put it. This meant that, following a confrontation with the authorities, A. L. was now verifying that the interior of a prison cell is not in the same connected component as the exterior of the prison cell. Gy Szekeres, a friend of Erdős from those times, claims that this is the context in which he first heard about the Jordan curve theorem.

Marin Mersenne (1588–1648) defined a Mersenne number to be an integer of the form $2^k - 1$ for k a positive integer. He conjectured that all such numbers with prime exponent are prime. For example,

$$2^2 - 1 = 3,\ 2^3 - 1 = 7,\ 2^5 - 1 = 31,\ 2^7 - 1 = 127$$

are all prime. It was F. N. Cole (1861–1927) (after whom the AMS Cole Prize in algebra is now named) who proved in 1903 that $2^{67} - 1$, the 67th Mersenne number, was not prime. At the October meeting of the American Mathematical Society that year, Cole gave a talk entitled "On the factorisation of large numbers." In it, he walked to the blackboard without uttering a word, calculated by hand the value of $2^{67} - 1$, calculated by hand the product of 193707721 and 761838257287 and let the audience observe that they are equal. He walked silently back to his seat. It is said that this is the first talk at an AMS meeting where the audience applauded. Cole spent all day every Sunday for three years doing the calculations necessary for this result.

It is said that in a certain year (in the early days of his career, when he served as a professor) one of Einstein's students came to him and said, "Professor Einstein, the questions on this year's exam are the same as last year's!" "True," said Einstein. "But this year all the answers are different."

Many years ago, there was a PDE conference in a small town in Italy. C.B. Morrey (1907–1984), Louis Nirenberg (1925–), and many other luminaries were in attendance. The mayor of this little town made a big fuss over the event, and an article appeared in the local newspaper. It featured a photo of Morrey lecturing, and the caption said, "Prof. L. Nirenberg giving a lecture on his research." Morrey looked at it and said, "That's not Nirenberg. Those are *my* equations."

John Horton Conway spoke at a recent event honoring H.S.M. Coxeter on his ninety-fifth birthday. Conway allowed that Coxeter had been an inspiration to him even very early in life: as a teenager, Conway had corresponded with Coxeter on problems of mutual interest. However, cautioned Conway, Coxeter was not always a positive influence. Once he was thinking about a Coxeter problem while crossing the street, and he was hit by a garbage truck.

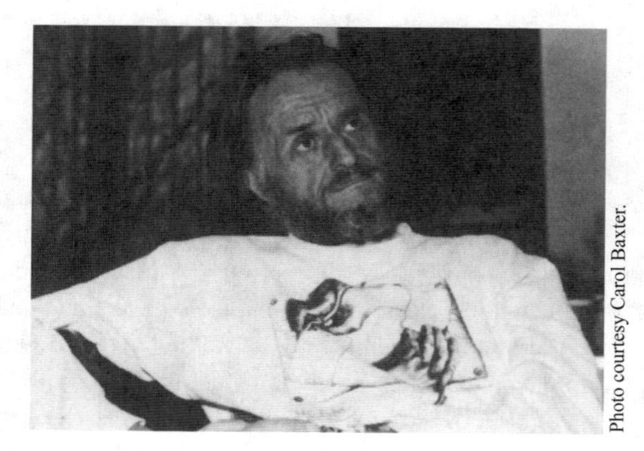

John Horton Conway

Photo courtesy Carol Baxter.

Great Pranks

In the district in India where the Tata Institute is located, it is illegal to buy liquor without a doctor's prescription. The custom then is to make an appointment with the right physician. The doctor says, "What is your problem?" The patient will reply "My elbow hurts" (or some such thing) and then the doctor will prescribe a bottle of gin. A visiting mathematician at the Tata Institute, unaware of this practice, made a doctor's appointment. He sat down, and the doctor said, "What is your problem?" The mathematician replied, "I have a sheaf of rings and an abelian functor, ..."

Antoni Zygmund was one of the many intellectuals who left Europe during the Nazi reign of terror. He was a Polish analyst and taught for many years at the University of Chicago. He always made a special effort to learn American ways, and availed himself of every opportunity to acquaint himself with local customs. One day he was wandering through a student lounge and found an excited group of students listening to the radio. He asked what was going on. They told him that they were listening to the World Series. He asked what that was. They said it was baseball. He asked what that was. So they explained to him about the American pastime, the infield fly rule, the designated hitter, and all the rest. Then they went on to tell that the season had 150 games, and it was topped off by the World Series. Zygmund thought a moment and then he said, "I think it should be called 'The World Sequence.'"

Marc Kac (1914–1984) was a mathematical physicist from Poland. When he first came to the United States, his English was minimal (he later became

Antoni Zygmund

quite fluent). He had a particularly hard time ordering anything to eat. But he did find a lunch counter in New York City where he could get his wishes met. He would walk in each day and ask for "a cream cheese sandwich and a Coke." One day a new counterman responded to this order by asking, "On toast?" Of course Kac was baffled, and responded with a bewildered look. After he left, he went home and got out his phrase book. Opposite "toast" he found "Gentlemen, the King!" Kac concluded that "toast" was some enthusiastic and friendly greeting among comrades. The next time he went to the lunchroom and placed his order, and the counterman replied, "On toast?" Kac shot his fist in the air and cried, "On toast!!"

When Kac taught at Cornell, he was on the Ph.D. committee of a very weak candidate. After the student had botched several questions, Kac endeavored to be kind and posed this very simple task: Describe the behavior of the function $f(z) = 1/z$ in the complex plane. The reply was, "The function is analytic, sir, in the whole plane except at $z = 0$ where it has a singularity." Of course this was perfectly correct, but Kac wanted more. "What is the singularity called?" he demanded. The student was stopped cold. Kac said, "Look at me. What am I?" The student brightened up and said, "A simple Pole, sir." Which was the correct answer.

Peter Farkas, now an applied mathematician, originally came to this country from Romania to study several complex variables with Raghavan Narasimhan at the University of Chicago. Farkas's English is now fluent, but it was only nascent when he first set foot in Chicago. One day he was in a restaurant struggling to make sense of the menu: such solecisms as "Monte Christo sandwich," "three pigs in a blanket," and "Reuben sandwich" were defying all his efforts with a dictionary. Finally Peter found a culinary item that he understood fully. With a big smile, he turned to the waitress and ordered, "A bacon, lettuce, and tomato sandwich and a Coca Cola." With a squint in her eye and a cigarette dangling from her lip, the waitress drawled, "Ya want mayo on your BLT?"

Marc Kac tells this story of Hugo Steinhaus (1887–1972): After the war, Steinhaus failed to attend a meeting of the Polish Academy. The secretary of the Academy wrote to ask him "how he justified his absence." Steinhaus replied, "As long as there are members who need to justify their presence, I do not have to justify my absence."

That there is a link between math and music is well known. Both UCLA and Washington University have courses on "Mathematics and Music." A number of famous artists have advanced degrees in math. These include Carole King (1941–) and Art Garfunkel (1941–). In the 1960's, Garfunkel had to decide whether to pursue a Ph.D. in math or go on the road with Paul Simon. We all know what choice he made. The rock singer Madonna (Veronica Louise Ciccone, 1958–) was a math major at the University of Michigan. It is difficult to say how this study has informed her songs.

There is a famous photo of basketball superstar Michael Jordan with his high school math teacher. In the caption she is saying, "Go into math, Michael. That's where the money is."

In the late 1950's, Tom Lehrer (1928–) was a great cultural hero among academics. A graduate student in mathematics at Harvard University, he spent his evenings performing satirical songs in Boston nightclubs. Some of his creations, such as "The Vatican Rag," "Werner von Braun," "Poisoning Pigeons in the Park," and "The New Math" have become cult classics. Lehrer's recordings can still be obtained in music stores and on the Web. Lehrer abruptly quit performing in the early 1960's. He now splits his time between California and New York, and is affiliated with the University of California at Santa Cruz (my *alma mater*). His department is Music, not Math, and he teaches a course in "History of Music." Lehrer did get a Bachelor's and a Master's degree in mathematics. Although he spent sixteen sporadic years working on a Ph.D., he never achieved that goal. When asked by a journalist why he stopped writing his mischievous songs, Lehrer replied, "When Henry Kissinger can get the Nobel Peace Prize, what is there left for satire?"

In the mid-1930's, a cabal of French mathematicians was formed with the purpose of writing definitive texts in the basic subject areas of mathematics. They ultimately decided to publish their books under the *nom de plume* Nicolas Bourbaki. In fact the inspiration for their name was an obscure French general named Charles Denis Sauter Bourbaki. This general, so it is told, was once offered the chance to be King of Greece but (for unknown reasons) he declined the honor. Later, after suffering an embarrassing retreat in the Franco-Prussian War, Bourbaki tried to shoot himself in the head—but he missed. He was quite the buffoon, and the authors of these mathematical texts decided that he was the perfect foil for their purposes.

In fact the founding mathematicians in this group—André Weil, Jean Dieudonné (1906–1992), Jean Delsarte (1903–1968), Henri Cartan, Claude Chevalley, and some others—came from the tradition of the École Normale Supérieure. This is perhaps the most elite university in all of France, but it also has a long-standing tradition of practical joking. Weil himself tells of one particularly delightful story. In 1916, Paul Painlevé (1863–1933) was a young and extremely brilliant Professor at the Sorbonne. He was also an examiner for admission to the École Normale Supérieure. Each candidate for admission had to undergo a rigorous oral exam, and Painlevé was on the committee. So the candidates came early in the morning and stood around the hall outside the examination room awaiting their turn. On one particular day, some of the more advanced students of the École began to chat with the novices. They told the youngsters about the fine tradition of practical joking at the school. They said that one of the standard hoaxes was that some student would impersonate an examiner, and then ridicule and humiliate the student being examined. The students should be forewarned.

Armed with this information, one of the students went in to take the exam. He sat down before the extremely youthful-looking Painlevé and blurted out, "You can't put this over on me!" Painlevé, bewildered, replied, "What do you mean? What are you talking about?" So the candidate smirked and said, "Oh, I know the whole story, I understand the joke perfectly, you are an impostor." The student sat back with his arms folded and waited for a reply. And Painlevé said, "I'm Professor Painlevé, I'm the examiner, …"

Things went from bad to worse. Finally Painlevé had to go ask the Director of the École Normale to come in and vouch for him.

When André Weil used to tell this story, he would virtually collapse in hysterics.

When Ralph Boas was the Executive Editor of the *Mathematical Reviews*, he made the mistake of printing the opinion that Bourbaki does not actually exist—in an article for the *Encyclopædia Britannica*, no less. The *Encyclopædia* subsequently received a scalding letter, signed by Nicolas

Ralph Boas (center, rear) and friends

Bourbaki, in which he declared that he would not tolerate the notion that anyone might question his right to exist. To avenge himself on Boas, Bourbaki began to circulate the rumor that Ralph Boas did not exist. In fact Bourbaki claimed that "Boas" was really an acronym; the letters B.O.A.S. were actually a pseudonym for a group of the *Mathematical Reviews's* editors.

Of all the members of Bourbaki, perhaps the one who served most frequently as the scribe was Jean Dieudonné. Dieudonné was also a prolific mathematician and writer in his own right. There is one particular incident which serves to delineate the two lines of his work. Dieudonné once published a paper under Bourbaki's name, and it turned out that this paper had a mistake in it. Some time later, a paper was published entitled, "On an error of M. Bourbaki," and the signed author was Jean Dieudonné.

André Weil tells that, in 1934, he and Henri Cartan were constantly squabbling about how best to teach Stokes's theorem in their respective courses at Strasbourg. He says, "One winter day toward the end of 1934, I thought of a brilliant way of putting an end to my friend's [Cartan's] persistent questioning. We had several friends who were responsible for teaching the same topics in various universities. 'Why don't we get together and settle such matters once and for all, and you won't plague me with your questions any more?' Little did I know that at that moment Bourbaki was born."

Weil claims that the name "Bourbaki" has an even longer history. In the early 1920's, when they were students at the École Polytechnique, one of the older students donned a false beard and a strange accent and gave a much-advertised talk under the *nom de plume* of a fictitious Scandinavian name. The talk was balderdash from start to finish, and concluded with a "Bourbaki's theorem" which left the audience speechless. One of the École's students claimed afterward to have understood every word.

In later years, Weil was at a meeting in India and told his friend Kosambi the story of the incident at the École. Kosambi then used the name "Bourbaki" in a parody that he passed off as a contribution to the proceedings of some provincial academy. Soon the still-nascent Bourbaki group determined that this would be its name. Weil's wife Eveline became Bourbaki's godmother and baptized him Nicolas.

Weil concocted a biography of Bourbaki and impugned him to be of "Poldavian descent." Nicolas Bourbaki of Poldavia submitted a paper to the *Comptes-Rendus* to establish his *bona fides*. Élie Cartan (1869–1951) and André Weil exercised considerable political skill in getting the man recognized and the paper accepted. It turns out that "Poldavia" was another concoction of the practical jokers at the École Normale. Puckish students frequently wrote letters and gave speeches on behalf of the beleaguered Poldavians. One such demagogue gave a speech that ended by saying, "And thus I, the president of the Poldavian Parliament, live in exile, in such a state of poverty that I do not even own a pair of trousers." He climbed onto a chair and was seen to be in his undershorts.

In 1939, André Weil was living in Helsinki. On November 30 of that year, the Russians conducted the first bomb attack on Helsinki. Shortly after the incident, Weil was wandering around the wrong place at the wrong time; his squinty stare and obviously foreign attire brought him to the attention of the police, and he was arrested. A few days later the authorities conducted a search of Weil's apartment in his presence. They found

- Several rolls of stenotypewritten paper at the bottom of a closet; Weil claimed that these were the pages of a Balzac novel.
- A letter, in Russian, from Pontryagin. It was arranging a visit of Weil to Leningrad.
- A packet of calling cards belonging to Nicolas Bourbaki, member of the Royal Academy of Poldavia.
- Some copies of Bourbaki's daughter Betti's wedding invitations.

In all, this was an incriminating collection of evidence. Weil was slammed into prison for good.

A few days later, on December 3, Rolf Nevanlinna (1895–1980)—at that time a reserve colonel on the general staff of the Finnish Army—was dining with the chief of police. Over coffee, the chief allowed that, "Tomorrow we are executing a spy who claims to know you. Ordinarily I wouldn't have troubled you with such trivia, but since we're both here anyway, I'm glad to have the opportunity to consult you." "What is his name?" inquired Nevanlinna. "André Weil." You can imagine Nevanlinna's shock. But he maintained his composure and said, "I know him. Is it really necessary to execute him?" The police chief replied, "Well, what do you want us to do with him?" "Couldn't you just deport him?" asked Nevanlinna innocently. "Well, there's an idea; I hadn't thought of it," replied the chief of police. And so André Weil's fate was decided.

Later Weil returned to France and was jailed for draft evasion. He liked to say many years after that jail was the perfect place to do mathematics: it was quiet, and there were few interruptions.

The movie *A Beautiful Mind*, about John Nash, has put mathematics on the map—at least in the popular cognizance. The movie depicts a ceremony, supposedly of long-standing at Princeton, of mathematicians showing their appreciation for another mathematician's theorem-proving ability by laying their pens before him on the Fine Hall Commons Room table. I never witnessed such a spectacle while at Princeton, nor did I hear of one. However, I inquired. One of my old professors assured me that there has never been any such tradition. But someone at Harvard had phoned him up, allowed that he saw this in the movie, and thought it was a great idea. They were going to start something up at Harvard.

My old teacher also told me of another eccentric who used to people the Fine Hall Commons Room. This fellow told people that he had installed a "people catcher" on the front of his car (something like a "cow catcher" on a train) because he was worried that he would run into Einstein (who was long deceased).

One of the most famous papers in all of mathematical analysis is "Maximal ideals in an algebra of bounded functions," authored by one I. J. Schark. It turns out that I. J. Schark never existed. The story is this. In the late 1950's, several top analysts were in residence at the Institute for Advanced Study in Princeton. At lunch one day, a vigorous discussion was held in which many of the foundational ideas of what was to become the theory of function algebras were developed. Everyone was quite pleased with the results of the discussion, but nobody wanted to take credit for them. So a name was cooked up from the initials of those present: The "I" comes from Irving Glicksberg, the "H" from Ken Hoffman, the "R" from Richard Arens, and so forth. A paper was written and I. J. Schark was designated as the author. The paper appeared in 1961 in the *Journal of Mathematics and Mechanics* and is one of the most frequently cited works in the literature. Kenneth Hoffman (one of the participants in the I. J. Schark *canard*) wrote a subsequent paper entitled "A note on the paper of I. J. Schark."

In a puckish mood, while I was a graduate student, I had a T-shirt made up that said "Legalize Cohomology." I wore it all over Princeton. In the math department it elicited laughter (from the students), raised eyebrows (from the faculty), and bewilderment (from the staff). On Nassau Street, the main thoroughfare of the rather staid and traditional town of Princeton, it seemed to engender looks of shock and horror. One day my fellow student Len Grissom showed up with a T-shirt that said "Kill All Homotopy Groups." That seemed to say it all.

At a certain university in Hungary—following on the sterling example set by Paul Erdős—professors worked hard to train their students for the big math competitions. In one particular year, they had a star student on whom they really pinned their hopes. On the day of a big exam, they sent him off to battle, expecting great glory for their math department. Indeed, he performed very well. On one particular problem, he got the correct answer of $\sqrt{2}+\sqrt{3}$. But he did not stop there. He wrote

$$\sqrt{2}+\sqrt{3} \approx 1.41...+1.73... \approx 3.14... = \pi.$$

The professors grading the exam were nonplussed. The student got the correct answer, and then went on to write a stream of charming sophistry. What to do? They debated matters long and hard, and in the end came down in the vein of conservative Hungarian tradition: the student received 0 points.

Wilhelm Stoll (1923–) is a distinguished complex analyst. He is perhaps the world's foremost expert on the Nevanlinna theory of holomorphic functions of several complex variables. One day he was lecturing on his favorite topic at a conference at the Mathematisches Forschungsinstitut Oberwolfach in the Black Forest of Germany. His title was "Value distribution theory with moving targets" and the lecture room was packed. Stoll spent half an hour preceding his talk recording several particularly large and complex formulae on the board. So the talk began with this awesome spectacle facing the audience. The presentation was exceedingly technical, and it soon became clear that even the most erudite *illuminati* were hopelessly lost. Stoll became frustrated to be lecturing to a sea of blank faces—especially since these were the people in the world most likely to understand what he was talking about. So finally, in his thick German accent, he said, "Look. It is like when you see fifteen birdies on a wire. And you take out your shotgun and you shoot two of them. How many are left?" He looked around the room expectantly. "You probably think 'thirteen,' right? Fifteen minus two is thirteen. But, no. When you shoot your gun, all the birdies fly away!" This was the high point of the conference.

According to legend, Solomon Lefschetz (1884–1972) was trained to be an engineer. This was in the days, one hundred years ago, when engineering was part carpentry, part alchemy, and part luck (the pre-von Kármán era). In any event, Lefschetz had had the misfortune to lose both his hands in a laboratory accident. This mishap was lucky for us, for he subsequently, at the age of 36, became a mathematician.

Lefschetz had two prostheses in place of his hands—they looked like hands, loosely clenched, but they did not move or function in any way. Over each he wore a leather glove. A friend of mine was a graduate student of Lefschetz; he tells me that one of his daily duties was to push a piece of chalk into Lefschetz's hand each morning and to remove it at the end of the day.

Lefschetz starred in one of the many films that the Mathematical Association of America made in the early 1960's featuring notable mathematicians speaking of their work. He gave a lovely intuitive lecture, punctuated by a cacophony of squeaky chalk, about his celebrated fixed point theorem. His feelings about the film were mixed; at one point he says on film, "I hope this is clear; it's probably about as clear as mud." After his lecture comes a filmed round-table discussion including John Moore, Lefschetz, and a few others. For ten or fifteen minutes they reminisce about the old days at Princeton. One person reminds Lefschetz that, in the late 1940's, during the heyday of the development of algebraic topology, they were on a train together. Lefschetz was asked the difference between algebra and topology. He is reported to have said, "If it's just turning the crank, then it's algebra; but if there is an idea present then it's topology." When Lefschetz was reminded of this story he became most embarrassed and said, "I couldn't have said anything like that."

Solomon Lefschetz

With his artificial hands, Solomon Lefschetz could not operate a doorknob, so his office door was equipped with a lever. Presumably he had difficulty with other routine daily procedures—dialing a phone, turning on a light, etc. By the time I was a graduate student at Princeton, Lefschetz was 87 years old. He was still mathematically sharp; indeed it was said that he was working on certain nonlinear differential equations that scared everyone else to death. But, in his later years, Lefschetz certainly had trouble getting around. In those days, Fine Hall, the mathematics building in Princeton, was having constant trouble with the elevators: push the button for the fifth floor and you are shot to the penthouse, down to the basement, and ejected on floor seven. There were several variations on this scenario, and the receptionist kept a log of complaints so that she could report them to the elevator repairman. [It might be noted that the state of the elevator was attributable in no small part to the student proclivity to use them for various pranks. More on this elsewhere in the book.] One day Lefschetz got into the elevator and it delivered him to the fourth floor "machine room." This room houses the air conditioning and heating equipment and is ordinarily only accessibly with a janitor's key. People are not supposed to go there, and ordinarily cannot. But with the elevator malfunctioning, Lefschetz was delivered to this forbidden floor. Poor Lefschetz wandered unwittingly out into the room, only to have the elevator doors shut behind him before he realized what was going on. He was trapped in total darkness, could not summon the elevator (no key), could not turn the doorknob to use the stairwell (no hands), and could not find a telephone (which, even had he found one, he probably could not dial). The members of the mathematics department rode that elevator for several hours, not realizing that Lefschetz was missing, before someone finally heard his desperate shouts and understood what had happened. Fortunately, Lefschetz was rescued and came away unharmed.

Speaking of the elevators at Princeton, one of my earliest memories as a graduate student is of the elevator emergency stop alarm going off three or four times per day. Especially puzzling was that everyone ignored it, and frequently they giggled. Bear in mind that this alarm only sounds if someone inside the elevator sets it off. It is sometimes used by the janitorial staff to hold the elevator at a certain floor; but the janitors never used it during the day. After I had asked around for some time, somebody finally took pity on me and told me the secret. When the mathematics department moved from old Fine Hall (which now houses Near Eastern Languages) to new Fine Hall (sometimes called "Finer Hall," overlooking "Steenrod Square"), Ralph Fox (1913–1973), the famous topologist, was annoyed that there was no men's room on his floor. So, whenever he had to use the facilities, he would take the elevator to the next floor, set the emergency stop alarm, do what needed to be done, and then return to his floor. Now I knew why everyone smiled when the alarm went off.

The mathematicians helped to design new Fine Hall, and they had the building built for their own convenience. There are two long, broad floors with classrooms and administrative offices. This is surmounted by the ten-floor tower with the faculty offices. Nine of the floors in the tower have faculty offices and a seminar room—nothing more. The floors are all identical. The entire impressive structure is surmounted by the thirteenth floor "Professors' Lounge," with a panoramic view of Princeton. The idea of the design is that the noise and hubbub of classes would not disturb the professors while they work. And students are not likely to be wandering the halls up in the tower. New Fine Hall is the tallest building on the Princeton campus—thirteen floors. When I was a student, it was one of the few buildings with an elevator. The students at Princeton tend to be rather bright, and they had good fun with the elevator. One evening they snuck into the building, armed with screwdrivers and other weapons of destruction, and rewired the elevator so that if you punched the button for floor n then you were taken to floor k, where k is a random function of n. Then they switched the signs from floor to floor: they took the office numbers and professors' names from floor n and moved them to floor k, for each pair of values n and k. Imagine the following morning: each of fifty or so sleepy faculty wandered into new Fine Hall, pushed the familiar button for his floor, went to his familiar office door (or so he thought) and found that his key would not fit. Of course there was no way for the professor to tell that he was on the wrong floor—short of looking out the window, and who would think of that? Soon the entire Princeton math faculty had converged on the departmental office, demanding to know why the lock was changed on his office door.

The new Fine Hall was dedicated in 1969 with much pomp and ceremony. Some wags immediately named it "Finer Hall." There is a large, abstract steel sculpture out in the square (which was soon named "Steenrod Square"), and it is festooned with many large pointy protrusions that reach for the sky. When it was being erected, one of the workers was accidentally dropped from a crane onto one of the pointy extremities of the sculpture. He was impaled there for most of the afternoon before rescue was effected, but by that time he had died. Years later there was a big lawsuit in connection with this incident.

The "roasting" of an individual is a peculiarly American custom: A group of close friends holds a fancy dinner in honor of the victim, after which they stand up one by one and make a collection of (humorously delivered) insulting remarks about him. Some anecdotes are in the nature of a roast. Here is an example. In the 1950's, it was said in Princeton that there were four definitions of the word "obvious." If something is obvious in the sense of Beckenbach, then it is true and you can see it immediately. If something is obvious in the sense of Chevalley, then it is true and it will take you several weeks to see it. If something is obvious in the sense of Bochner, then it is false and it will take you several weeks to see it. If something is obvious in the sense of Lefschetz, then it is false and you can see it immediately.

In 1978, I was once standing in line with Edwin Beckenbach (1906–1982) at the UCLA Math Department's fancy new photocopy machine. This contraption had every convenience—sorters, duplex copying, and so forth. Edwin and I chatted, and he mused about the days when he was department chairman in the late 1940's. "You know, the biggest fight I ever had with my department was over whether to replace our hand-cranked mimeograph machine with a motorized one. I was in favor of the modern technology, but the department felt that the hand-cranked machine was adequate." Beckenbach smiled and pointed out that he won the battle.

Angus Macintyre (1941–) was a logician at Yale. One day he got an offer from Oxford University in England. This was his "dream job," and he seriously considered accepting it. Yale has a great tradition in logic (going back to the days of Abraham Robinson and beyond) and they did not want to lose Macintyre. His colleague George Seligman (1927–) said, "We had better meet Oxford's offer. Cut Macintyre's salary in half."

It is said that one day Kakutani was walking down the hall at Yale and ran into a student. The student said, "Professor Kakutani, can I come to your office at 4:00 p.m. today?" Kakutani readily said "Yes," and began to go on his way. The student called over his shoulder, "And will you be there, Professor Kakutani?" "No," came the easy reply.

Norbert Wiener once said that he wanted to write a poem about a girl from Walla Walla who went to Pago Pago to dance the Hula Hula. The puckish Ralph Boas did it for him (with apologies to the mavens of political correctness). Ralph gave me the poem, entitled "Echolalia," together with his permission to publish it:

Guido Weiss

Lulu from Walla Walla was a devotee of dance;
She danced a wicked can-can in a tutu sent from France.
She said, "I'm going gaga; in toto, it's a bore;"
So she went to Bora Bora and to Pago Pago's shore,
Where she studied hula hula and worked for an A. A.,
But her work was only so-so and they wouldn't let her stay.
Now she's gone to Baden Baden, to a go-go cabaret;
Billed as Lulu in the muu-muu, she's performing every day.
When she lets the muu-muu drop, the old folks
nearly drop their teeth
When they see the lava-lava that is all she's got beneath.

In the academic year 1968–1969, Guido Weiss (my colleague) spent a one-year sabbatical in Princeton. He and E. M. Stein had an intense collaboration during that time; it resulted in their masterpiece *Introduction to Fourier Analysis on Euclidean Spaces*. This all occurred while Stein was chairman of the math department at Princeton. When I was a graduate student there, I took a delightful course on semi-groups of operators from Ed Nelson (1932–). He frequently told us stories of his colleagues, and one concerned a dinner party that he and his wife gave in 1969 that included Stein and Weiss as guests. At one point Weiss allowed that in a few weeks he and his family were to take a much-needed vacation. Stein interjected that Weiss couldn't do that; he was supposed to be finishing Chapter 9 of "the book." Everyone was rather non-plussed, and the conversation went in another direction. Next day, Nelson's son came up to him and said, "Dad, will Professor Stein let us take a vacation this year?"

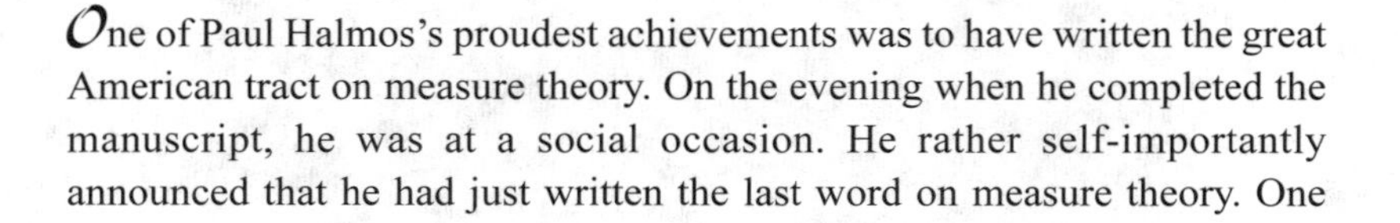

One of Paul Halmos's proudest achievements was to have written the great American tract on measure theory. On the evening when he completed the manuscript, he was at a social occasion. He rather self-importantly announced that he had just written the last word on measure theory. One

Paul Halmos and Pizzicato

puckish individual asked, "Oh? And just what is that last word?" Halmos could not remember, so he rushed back to his office to check. Then he rushed back to tell. The last word in measure theory was "*X*."

Halmos always tried to take an original approach to his teaching. One day, in a class on vector spaces, he put on the blackboard a list of six things that a vector is *not*. For example, it is not **(i)** an n-tuple of numbers, **(ii)** an arrow in the plane, and so forth. One of his students, a feisty middle-aged ex-dropout, whom Halmos calls Wilhelmina (though that was not her real name) took exception to this approach. She looked up "vector" in the *Encyclopædia Britannica* and then filed a complaint with the appropriate Associate Dean, claiming that Halmos was narrow and prejudiced. The Dean had been around the block a few times, so he did not take the complaint very seriously. Still, he took a few moments to chat with Halmos and to satisfy himself that there was nothing really amiss. Then Halmos went back to the teaching of vector spaces. On the final exam for the course, Halmos put the standard question, "Is the set of complex numbers a vector space over the real numbers?" Wilhelmina's answer was, "That's a trick question, and if there's one thing I learned in this course, it's not to answer trick questions."

A popular story about Edward Titchmarsh (1899–1963) is that he once announced a course that would be a "double series" of lectures—lasting not one academic year but two. The first half of this marathon course ended in April, and there was a six month break before the next half began the following October. Titchmarsh strode into the room that October, picked up a piece of chalk, and said, "Hence, …."

During the early 1970's at Princeton, Eric Friedlander was an Assistant Professor in the Princeton mathematics department. One semester he was teaching an advanced course in his specialty—algebraic topology. On a particular day he said to the class, "After today's class you will have heard all my ideas. After next class you will have seen all my shirts."

Richard Courant was a student of Hilbert in Göttingen. One day he noticed that the zipper on Hilbert's fly was broken. In those days the separation between professor and student was immense: students were deferential and professors were lordly. He was at a loss to know how to approach the important professor and tell him of this embarrassing development.

Courant continued to ruminate on the problem. As luck would have it, the weekly departmental bicycle ride was that afternoon, and both Hilbert and Courant went. Hilbert had a mishap, and fell off his bicycle. A quick thinker, Courant ran up to Hilbert, confirmed that he was all right, and exclaimed, "Professor Hilbert! I am glad to see that you are unhurt. But oh, look, your fly is broken!" Hilbert replied, "Courant, you idiot, it's been broken all week."

Hans Grauert is one of the most distinguished mathematicians of our time. Many of the seminal ideas in several complex variables, including a very powerful solution of the Levi problem, are due to his efforts. Grauert is a rather distant and austere individual, and usually can be seen dressed all in grey and black, with a very serious look on his face. Grauert was a student of Behnke many years ago. One of Grauert's duties was to accompany his teacher whenever the great man had to move his bowels.

One of the most famous Wiener stories, for which I have independent confirmation suggesting that (in some form) it is true, is about the day that the Wiener family was moving to a new house. At breakfast, Wiener's wife made a grand ceremony of taking his old house key and putting the new house key on his key ring. She wrote the new address (still within walking distance of the university) on a slip of paper and carefully instructed Norbert that today he was to go home to the new abode. All to no avail.

Because Norbert, in the middle of the day, used the slip of paper to answer a math query for someone. So he no longer had the address of his new home. At the end of the day, a creature of habit, he moseyed back to his usual home, only to find nobody there. Worse, his key would not fit in the lock. Looking in the window, he could see that all his possessions were missing. Going into a panic, Wiener proceeded to wring his hands and run around in the yard.

Presently, a little girl came down the street. Wiener ran up to her and said, "Little girl, you don't know me. But I'm in distress. My family has

G. H. Hardy

disappeared, my key won't fit in the lock, and my house is empty." "Yes," replied the little girl. "Mommy sent me for you." [In fact one wag asked Wiener's daughter, years later, if this story were true. The reply was, "Not quite. He would have recognized his daughter."]

Wiener was in the habit of reading the newspaper at the same time each day in one of the M.I.T. lounges. The mischievous students would sneak in and set the bottom of Wiener's newspaper on fire, with dramatic effect. This was a joke that they could apply repeatedly—and they did so with gusto.

Stefan Bergman had a self-conscious sense of humor and a loud laugh. He once walked into a secretary's office and, while he spoke to her, inadvertently stood on her white glove that had fallen on the floor. After a bit she said, "Professor Bergman, you're standing on my glove." He acted embarrassed and exclaimed, "Oh, I thought it was a mousy." [It should be mentioned here that there are a number of wildly exaggerated versions of this story in circulation. But I got this version from the secretary in question.]

G. H. Hardy was a mathematical analyst who wrote many papers—more than 100 of them with his friend J. E. Littlewood. His favorite problem was undoubtedly the Riemann hypothesis. Hardy regularly left England and went to the continent for his summer vacation. There he would visit with his friend Harald Bohr (1887–1951). They had a fixed routine. First they sat down and talked, and then they would go for a walk. As they sat, they wrote down an agenda. The first point on the agenda was always the same: "Prove the Riemann hypothesis." This agenda item was never successfully addressed, but Hardy always insisted that it go on the agenda.

Hardy felt that he had a personal conflict with God: God was his enemy, and would go out of His way to make life miserable for Hardy. For example, Hardy would always bring an umbrella to his beloved cricket matches, for he was sure that God would not give him the satisfaction of being prepared for rain.

One year Hardy had to return in some haste from Denmark to England. Only a very small boat was available. The North Sea is notoriously treacherous, and the chances of such a boat sinking were serious. Nervous as he was, Hardy *had* to go. He sent a postcard to Bohr saying, "I have proved the Riemann hypothesis." Hardy was confident that God would not grant him the honor and prestige of dying and leaving people with the idea that G. H. Hardy had taken a proof of the Riemann hypothesis to his watery grave.

It is said that Hardy had four ambitions in life:

- To prove the Riemann hypothesis;
- To score the winning play in an important game of cricket;
- To murder Mussolini;
- To prove the non-existence of God.

So far as we know, Hardy did not achieve any of these goals.

This story is not about mathematics or mathematicians *per se*, but it is so good that we cannot omit it from this collection. In 1948, the noted physicists Hans Bethe (1906–) and George Gamow (1904–1968) were writing a paper together. Noting the analogy with the sequence of Greek letters α, β, γ, they regretted the lack of a "first character" in their collaborative nameset. They conscripted former student Ralph Alpher to be a co-author. So now there is a celebrated paper in *The Physical Review* by Alpher, Bethe, and Gamow entitled "The Origin of Chemical Elements." The theory expounded therein has important consequences for the Big Bang Theory of the origin of the universe.

Once von Neumann and Ulam were having verbal jousts about the quality of all the brain power assembled at Princeton University and the Institute for Advanced Study. Ulam suggested that the situation was similar to the division of rackets among Chicago gangsters. The "topology racket" was probably worth five million dollars; the "calculus of variations racket" another five million. "No," said von Neumann. "That is worth only one million."

A prominent British mathematician was asked, some years ago, who the three best mathematicians in Britain were. Without hesitation he responded, "Hardy, Littlewood, and Hardy–Littlewood."

Littlewood once wrote a notice for the Ballistic Office. It ended with the sentence, "Thus σ should be made as small as possible." This statement did not in fact appear in the typeset document. P. J. Grigg then asked Littlewood, "What is that?" A speck in a blank space at the end of the notice turned out to be the tiniest σ that anyone had ever seen.

After World War I, Hardy wrote to Landau saying that he had not been a fanatical anti-German, and that he was confident that Landau would wish to resume former relations. Landau replied, "As a matter of fact my opinions were much the same as yours, with trivial changes of sign."

A famous remark of Bertrand Russell is that the Ten Commandments should be headed (like an examination paper) "Not more than six to be attempted."

The University of California at Santa Cruz mathematics department (my *alma mater*) once spawned a scheme to beat the roulette wheel in Las Vegas. Major players in the scheme were J. Doyne Farmer (1953–) and Norman Packard (1956–). Ralph Abraham (1936–) played an ancillary role, and so did several others of the mathematics faculty. The full story of this adventure is chronicled in the book *The Eudaemonic Pie* by Thomas Bass.

The gist of the scheme was predicated on the observation that any roulette wheel is imbalanced in some small way. The numbers do not come up in a completely random manner. Therefore, if one makes a sufficient number of observations of spins of a given roulette wheel, then one will have enough information to predict future spins—at least with a probability of success that is greater than purely random guessing. Bear in mind that the work being described here was done in the late 1970's, the dawn of the microcomputer age. What the participants did was to design a microcomputer that could be hidden in the heel of a shoe. A team of two "investigators" would go to Las Vegas to implement the technology. Investigator *I* would record spins on the chosen roulette wheel and transmit them by radio to Investigator *II*, who had the computer concealed in his shoe. The computer would make the necessary calculations and then prepare itself to make predictions of future spins. At the appropriate moment, when enough data had been gathered, Investigator *II* would begin to bet.

These merry pranksters actually had financial backers for their scheme. And they indeed made money, but at a rate so slow that the financial backers eventually withdrew their support. In the end, the fatal flaw in the project was that Investigator *II* sweated so heavily into his shoes that he shorted out the computer. So much for modern technology.

Washington University math graduate student Marcus Feldman was enjoying Paris on the left bank, wandering the Boulevard St. Michel and its environs. This was in the late 1960's, and Paris was bustling with life. Mark had his guitar over his shoulder and was enjoying the day. All at once, he was confronted by three tough-looking guys, each with several days growth of beard and sporting black leather jackets, dark glasses, and motorcycle chains for belts on their greasy jeans. They spoke guttural French, which Mark could hardly understand. But the three tough guys, through grunts and gestures, made it clear that they wanted Mark's guitar. This was Mark's prized possession, and he hardly wanted to relinquish it. But the alternative seemed to be mayhem, so after several moments' repartee he reluctantly handed over the instrument. The three hard guys instantly jumped into a trio formation and sang "One for the money, two for the show, three to get ready now go, man, go...."[1]

John von Neumann was an amazing mathematician and an amazing calculator. He also had a photographic memory: he could effortlessly recite long

[1] These are the opening lyrics to *Don't Step on My Blue Suede Shoes* by Carl Perkins, a famous song from the roots of rock and roll.

passages from novels that he had read twenty years before. He of course played an instrumental role in developing one of the first stored-program computers at the Institute for Advanced Study in Princeton. In those days, von Neumann was extremely active as a consultant. He was constantly coming and going, all over the country, to government agencies and companies, giving out the benefit of his erudition. It was said that his income from these activities (on top of his princely Institute salary) was quite substantial. And he was already rather wealthy with family money.

During one of von Neumann's consulting trips, Herman Goldstine (1913–) and the others working on the new computer got it up and running for a test. They fed it a large amount of data from meteorological observations, ran it all night, and came up with very interesting solutions in the morning. Later that day, von Neumann returned from his trip. Wanting to pull a prank on Johnny von Neumann, they decided *not* to tell him that they had the computer up and running, but instead to present their results as though they had obtained them by hand. At tea, they told von Neumann that they had been working on such and such a problem, with thus and so data, and in the first case had come up with .... "No, no," said von Neumann. He put his hand to his forehead, threw his head back, and in a few moments gave them the answer. It was the same answer that the machine had generated. Then they said, "Well, in the second case we got .... "No, no," said von Neumann. "Let me think." He threw his head back—it took longer this time—but after several moments he came up with the answer. Finally his collaborators said, "Now in the third case ...." Again, von Neumann insisted on doing the calculation himself. He threw his head back and thought and thought and thought. After several minutes he was still thinking and they blurted out the answer. John von Neumann came out of his trance and said, "Yes, that's it. How did you get there before I did?"

During World War II, Paul Halmos worked for a time at the Radiation Laboratory in Cambridge, Massachusetts. At one point they came up with an intractable problem. They could not answer it, but they knew just the guy who could. Unfortunately he did not have the necessary security clearance. After considerable thought, they were able to transform their problem into something else in which the original "defense application" was no longer discernible. Surely it would be fine to share this new problem with the expert who had no security clearance. And so they did. It was a hard problem, and the expert did not solve it in a day, or even in a week. But after ten days he did solve the problem. He phoned up Halmos's team in some excitement and told them of his achievement. They drove down with increasing anticipation to hear the answer. He had in fact transformed the question back to the original "top secret" problem and then he had solved it.

The British scientist Isaac Newton and the mathematician John Wallis (1616–1703) were friends. According to his diary, Newton once bragged to Wallis about his little dog Diamond. "My dog Diamond knows some mathematics. Today he proved two theorems before lunch."

"Your dog must be a genius," replied Wallis.

"Oh, I wouldn't go that far," offered Newton. "The first theorem had an error and the second had a pathological exception."

Great People

Archimedes (287 B.C.–212 B.C.) was devoted to his mathematics, and the lack of a convenient writing surface was hardly sufficient to hinder his efforts. After leaving his bath he would anoint his body with olive oil, and then trace his calculations with a fingernail on the oily skin.

Archimedes was the father of all mathematics, and one of the greatest mathematicians of them all. When invading Roman troops marauded Syracuse, where Archimedes was living, he contented himself with his mathematics. Marcellus (268 B.C.–208 B.C.), the general who commanded the conquering troops, ordered that Archimedes should be protected. Archimedes's first intimation that the city had been sacked was the shadow of a Roman soldier falling across his diagram in the soil. One version of the story is that the heathen stepped on Archimedes's diagram, causing the mild-mannered scholar to become angry and exclaim, "Don't disturb my circles!" Enraged, the soldier drew his sword and slew Archimedes.

Isaac Barrow (1630–1677) was Isaac Newton's teacher. Legend has it that he gave up the Lucasian Chair of Mathematics so that Newton could have a suitable post (although other versions of the story indicate that Barrow had another attractive option awaiting him). Barrow was no favorite of King Charles's protégé the Earl of Rochester, and vice versa. One day the two encountered each other at court. The ensuing conversation proceeded thusly:

The Earl: "Doctor, I am yours to my shoe-tie."
Barrow: "My lord, I am yours to the ground."

Lars Ahlfors

> The Earl: "Doctor, I am yours to the center."
> Barrow: "My lord, I am yours to the antipodes."
> The Earl: "Doctor, I am yours to the lowest pit of hell."
> Barrow: "And there, my lord, I leave you."

In 1936 the first Fields Medals were awarded to Lars Ahlfors and Jesse Douglas. A few years later, when war overtook Europe, Ahlfors was in a desperate situation in which he had to buy a boat passage to ensure his safety, but he had insufficient funds. Ahlfors's account of the story is this:

> I can give one very definite benefit [of winning a Fields Medal]. When I was able to leave Finland to go to Sweden, I was not allowed to take more than 10 crowns with me, and I wanted to take a train to where my wife was waiting for me. So what did I do? I smuggled out my Fields Medal, and I pawned it in the pawn shop and got enough money. I had no other way, no other way at all. And I'm sure it's the only Fields Medal that has been in a pawn shop… As soon as I got a little money, some people in Switzerland helped me to retrieve it.

Ahlfors later quipped that this was one of the few practical uses for a Fields Medal.

In 1964, Heisuke Hironaka won the Fields Medal for proving the celebrated resolution of singularities theorem for algebraic varieties. This was a big event, and of course the Japanese were thrilled that he would be accorded such a high honor. An audience was arranged for Hironaka with the Emperor of Japan. Part of the ceremony was that Hironaka was to give a (brief) lecture to the Emperor about what he had accomplished. Hironaka took this injunction very seriously, but he was painfully aware that the Emperor did not know any but grade school mathematics. He made a big effort to explain resolution of singularities in the most elementary terms, using the cusp of a cubic curve to illustrate the idea. He prepared elegant graphic displays to aid the Emperor's understanding. He was quite nervous that the whole thing would go well. On the appointed day, Hironaka showed up at court, was presented to the Emperor, and then he gave his lecture. It was only 20 minutes—short and sweet—and at the end Hironaka gave a sigh of relief. He felt that he had touched the high points, not gotten bogged down in details, and had given even the most rank amateur at least a rough idea of what resolution of singularities was about. As a courtesy, he bowed to the emperor and asked if there were any questions. The emperor smiled, raised his index finger, and said, "Just one. What about characteristic p?"

Descendents of Nikolai Ivanovich Lobachevsky (1793–1856) settled in Providence, Rhode Island. Their descendants, in turn, have opened a retail automobile business. It is called *Lorberg Motors*.

Two sons of Carl Friedrich Gauss, the prince of mathematics, settled in the St. Louis area. Their descendents still live in this midwestern metropolitan area. In fact a local automobile has a vanity license plate that says "GAUSS."

Vladimir Ezhov is a complex analyst who works in Adelaide, Australia. He is also a gifted tennis player, once ranked 25th in Moscow. In this capacity, he was Boris Yeltsin's tennis coach. He allows that Yeltsin cheated. When Yeltsin hit the ball out of bounds, his KGB security guards would muscle up to Ezhov and insist that the ball was in.

Marshall Stone was one of the most eminent mathematicians of the twentieth century. He played a seminal role in building up the University of Chicago Mathematics Department in the 1940's and early 1950's. He had a long and distinguished career, and in his later life was a statesman for modern mathematics. A few years ago a big conference was held at the University of Chicago to remember and to honor Stone's many contributions. It was aptly entitled "The Stone Age."

One of Marshall Stone's claims to fame is the "Stone–Weierstrass theorem," a deep and important generalization of the Weierstrass approximation theorem. This is the sort of result that could have been published in the *Annals of Mathematics*. But Stone sent it to *Mathematics Magazine,*[1] because he had promised them a paper to help them get off to a good start. And that is where this blockbuster paper appears.

When Stone was chairman of the University of Chicago Mathematics Department in the 1940's, he used to come to work early each day so he could go from room to room in Eckhart Hall and wash the blackboards. Such was life in the Stone age.

G. D. Birkhoff (1884–1944) was perhaps the dominant figure of American mathematics for much of the first part of the twentieth century. Some have said that he was the "first great American mathematician"—meaning that he was the first American-born mathematician whose work was taken seriously by the Europeans. Certainly Birkhoff's ergodic theorem was a milestone of its time. There are stories, based in part on Wiener's rendition in his autobiography, that Birkhoff was antisemitic. Wiener in fact claims that he could not get educated in the United States, nor land a job consonant with his talents, because of Birkhoff's pernicious influence. But there is evidence to the contrary. Constance Reid tells me of Birkhoff recommending a Jewish Ph.D. from Harvard to the mathematics chairman at Rochester. When Rochester balked, "He's Jewish, isn't he?" Birkhoff said, "If you don't take this one, you will never get another Harvard Ph.D."

---

[1] His article, entitled "The generalized Weierstrass approximation theorem," appeared in volume 21(1948), pp. 167–184.

Constance Reid

Constance Reid also provides tales of Richard Courant's group at NYU. She reports that it was remarkably congenial. The only really serious conflict was the one that arose between Courant and Lipman Bers—over a certain tenure case to which Bers was rather close; as a result of the heated argument, which could be heard late one day reverberating in the halls of the Courant Institute, Bers left NYU and moved to Columbia University. So much bitterness remained that even years later Bers was not willing to be interviewed by Reid for her book on Courant. After that book was published, K. O. Friedrichs asked Bers how he liked it. "It's a very good book. Very honest," Bers said—"But will they know how charming he was?"

Richard Arens was a gifted mathematician, but he had no interest in pretense or pomp. He could have taken a position at U. C. Berkeley, but instead chose to go to UCLA (which is a great mathematics department today, but was more modest in the 1950's). As a result, Arens was something of a campus celebrity. Pilgrims would come from all around, knock breathlessly on Arens's door, and say, "Professor, can I see you?" Arens would smile, spread his hands, and say, "Start looking!"

Hypatia of Alexandria (370–415) was the first great woman mathematician. She was also a philosopher, and arguably the first feminist. She was the daughter of Theon and wife of the philosopher Isidorus.

Hypatia is remembered for her commentaries on Diophantus's *Arithmetica* and Apollonius's *Conics*. She studied Euclid, and is credited with the version of Euclid's third book that we have today. Hypatia was loved and admired by the entire city of Alexandria, so much so that it made the archbishop Cyril jealous. It is said that he arranged to have her murdered by a mob of Christians. They flayed her with oyster shells, dragged her body through the streets, and mutilated her remains.

Some of Hypatia's wisdom:

> All formal dogmatic religions are fallacious and must never be accepted by self-respecting persons as final.
>
> Reserve your right to think, for even to think wrongly is better than not to think at all.
>
> To teach superstitions as truth is a most terrible thing.

Today some prominent mathematical families have named their daughters Hypatia.

When he was quite old, Abraham de Moivre (1667–1754) slept twenty hours per day. Right before his death, he asserted that it was necessary for him to sleep fifteen minutes more each day. The day after he reached a total

Photo courtesy of Randi D. Ruden.

Hypatia S. R. Krantz **Hypatia of Greece**

of over twenty-three hours, he then slept up to the limit of twenty-four hours and expired.

One of the great ideas in modern mathematics is Dennis Sullivan's theory of rational homotopy. When he was asked how he could have come up with such a marvelous idea, he said, "I read Hassler Whitney's book *Geometric Integration Theory*, and I really understood what he was talking about."

Luther Pfahler Eisenhart was for many years the grand old man of geometry at Princeton. He wrote the seminal book *An Introduction to Differential Geometry, With the Use of The Tensor Calculus*. There is a rather grand arch, named in his honor, at the Graduate College in Princeton. One day my friend Hung-Hsi Wu and another mathematician were driving out to the Institute for Advanced Study, and they drove under the arch. Wu stopped the car and said, "Eisenhart. All formulas, no ideas."

Walter Craig was an Assistant Professor at Stanford for several years. One evening he was at a colloquium dinner, and sitting next to the venerable Halsey Royden. At one point, late in the evening, Halsey turned to Walter and said, "Walter, do you have two dollars?" Walter allowed that he did, and handed the money over. "Congratulations," said Royden. "You just bought my car (a Fiat Mille Cinque Cente)." Next day Royden handed over the papers, and Walter Craig found himself saddled with a new, and suitably antiquated, vehicle.

Peter Gustav Dirichlet (1805–1859) married Rebecca Mendelssohn, sister of the composer Felix Mendelssohn. Felix and Rebecca were grandchildren of the philosopher Moses Mendelssohn. Dirichlet was against the writing of letters. He never wrote them, and he never sent them. Many of his friends lived out their lives without ever receiving correspondence from the noted mathematician. Dirichlet made an exception when his first child was born. He wired his father-in-law with the message "2 + 1 = 3."

Richard Dedekind

Richard Dedekind (1831–1916) was a student of Carl Friedrich Gauss in Göttingen. After he retired from his Professorship at Brunswick, he lived a very quiet life and saw few people.

As a result, the *Jahresbericht* of the German Mathematical Society inadvertently reported that Dedekind had died. It gave the day, month, and year of his demise. Dedekind took umbrage with this report. He wrote a letter to the editor saying, "In your communication, page so and so, of the *Jahresbericht*, concerning the date, at least the year is wrong."

When Don Marshall, currently chairman at the University of Washington in Seattle, was an undergraduate, he claims that he paid his way through school as follows. He drove a small sportscar, a Triumph TR-6, around Los Angeles. Since the car was small, other drivers—in their Cadillacs and Mercedes—frequently either didn't see the car or didn't make suitable allowances for it. So Don was often rear-ended. And the doors would fall off. Really. This was no big deal; Don could re-mount the doors in five minutes. But instead he would make a huge fuss, remonstrating that the errant driver had wrecked his lovely sports car. In typical Los Angeles fashion, the driver at fault would usually pay Don several hundred dollars on the spot to just shut up and go away.

John Fry is the successful and wealthy owner of *Fry's Electronics*, a chain of emporium-style electronics stores. He is currently building The American Institute of Mathematics, a splendid new math institute in Morgan Hill, California. The building will be modeled after the Alhambra in Grenada, Spain. There will be accommodations for 32 resident mathematicians, a dining hall with a first class chef, and a large and well-endowed library. In fact Fry has been collecting rare math books for many years. He has a copy of the first printed edition of Euclid, a copy of Napier's book of logarithms, first editions of Bolyai, Lobachevsky, and Fermat, and many other remarkable volumes. One of his real treasures is a first edition of Newton's *Principia.* Isaac Newton occupied the Lucasian Chair of Mathematics at Cambridge University, and the current occupant of that Chair is Stephen Hawking. So Fry asked Hawking to sign his copy of Newton. Hawking agreed, but in fact he cannot wield a pen. So he instead gave his thumbprint and then Hawking's wife wrote in the book attesting that this is the thumbprint of Stephen Hawking.

There are not many monuments in the United States to mathematicians or to mathematics. But Europe is littered with graves and memorials to celebrated mathematical scholars. A street in Torino, Italy is named after Faà di Bruno (1825–1888) and another is named after Beniamino Segré (1903–1977). Yet another street is Via Lagrangia, running into the Piazza Lagrangia. Oslo has a street named after Sophus Lie (1842–1899) and Germany has streets named after Gauss and others. [Actually, my home town of Redwood City, California has a street named after Euclid (c. 325 B.C.–265 B.C.). But this is not typical in the U.S. John Fry—see above—once sought to have all his stores on streets named after mathematicians. Indeed, two of them are on Euclid and Hamilton Streets (William Rowan Hamilton, 1805–1865). But, now that he has twenty-five stores in four states, this goal seems unattainable.] Two remarkable mathematical artifacts are these: In the Cité de Ville in Toulouse, France there is a room full of statuary. One of the larger statues is of a seated image of Pierre de Fermat. A stone sign—part of the sculpture—says "The Father of Differential Calculus." Seated in Fermat's lap is a scantily clad muse, showing her ample appreciation for Fermat's mental faculties. Second, on the grounds of the royal palace in Oslo there is a statue of Niels Henrik Abel (1802–1829). He is shown with his foot atop two prone and vanquished men. It has been conjectured that one of these is the fifth degree polynomial equation and the other an elliptic function.

The important publication *Mathematical Reviews* has been the venue for some rather delicious displays of opinion, prejudice, and vexation. Ralph Boas once wrote a review that said, "This paper fills a much-needed gap in the literature." Clifford Truesdell (1919–2000), in a remarkable fit of pique, once said, "This paper gives incorrect proofs of trivial theorems. The basic mistake, however, is not new."

Clifford Truesdell was a prolific writer, author of a great many papers and books. Even though he lived in the modern age, he did all his writing by candlelight with a quill (cut from a goose's feather). He also ate dinner each night wearing a dinner jacket, cummerbund, and tuxedo shirt; he said he would feel naked dining any other way. If one was invited to the Truesdell home, one was greeted in the foyer by the host and his wife, and also by a large mural of Salome with the head of John the Baptist on a platter. But a few small changes had been made. The head of John the Baptist was replaced by the head of Clifford Truesdell and that of Salome was replaced by the head of Mrs. Truesdell. [By some accounts there was also a portrait of Mrs. Truesdell, in the buff, hanging in their dining room.]

Norbert Wiener was well known to be eccentric. One of his FBI documents, obtainable through the Freedom of Information Act, reveals that Wiener was recognized both by the government and by MIT to be "emotionally unstable." Another government document characterizes Wiener as "harmless." It then goes on to recount a tale of Wiener driving to a conference in Pittsburgh, forgetting that he brought his car, and taking the train back home to Boston. Not finding his car at home, Wiener then reported it to be stolen.

Wiener attended a session of the Indian Science Congress at Hyderabad in India in 1954. Those were the bad old days of the McCarthy era, and everyone was afraid of appearing to consort with communists or communist sympathizers.

Wiener voluntarily gave a report to the FBI on his experiences at the conference in Hyderabad. He reported on "pug-uglies" from the Soviet Embassy policing the conference. He claimed that the Russians behaved very badly.

Wiener decided that it would be appropriate to engender sour relations with the Russians, and he proceeded to create bad blood. The following is a direct quotation from the FBI report:

> The professor [Wiener] said that he decided early during the Conference to try and needle the Russian delegates whenever an appropriate occasion arose. For instance, he told one that "Your newspaper, *Pravda*, has depicted me as a capitalist warmonger and a cigar-smoking capitalist, at that." On a second occasion, Professor Wiener, who speaks Chinese, told another Soviet delegate "You should really learn Chinese, you know. It is the language of a very important people." Professor Wiener jokingly referred to a woman Soviet delegate who wore a particularly drab and unattractive hat. The professor remarked that the delegates from other countries commented among themselves that the hat sat level on her head and that it was at least a proper hat politically speaking having neither a "right" nor a "left" deviation.

Solomon Lefschetz was famous for his overbearing self-confidence and, sometimes, his arrogance. He could intimidate most other mathematicians. At committee meetings he would pound his fist on the table with terrifying effect. It is even said that he sometimes pounded his shoe on the table (long before Nikita Krushchev made this technique famous). So it is with pleasurable surprise that one hears of exceptions. The one I have in mind is a certain unflappable graduate student at the time of the student's qualifying examination. The qualifying exam at Princeton is administered as one long oral exam: three professors and one graduate student are locked in an office for about three hours. The student is examined on real analysis, complex analysis, algebra, and two advanced topics of the student's choosing. Our confident student had Lefschetz on his committee. Lefschetz was known for, among other things, profound generalizations of Picard's theorems (Emile Picard, 1856–1941) in function theory to several complex variables. So it came as no surprise when Lefschetz asked the graduate student, "Can you prove Picard's Great Theorem?" Came the reply, "No, can you?" Lefschetz had to admit that he could not remember, and the exam moved on to another topic.

It is also said that Lefschetz slept through lectures but would always awake and ask a terrific question at the end. One day the lecturer got stuck in the middle of his talk. There were several minutes of silence while the poor fellow tried to rescue his theorem. This procedure threw off Lefschetz's rhythm. He woke up, looked around, and declaimed "That was a wonderful lecture." He then proceeded to ask one of his famous questions.

On another occasion, Wiener woke up in the middle of a lecture. He peered slowly at each of the blackboards, evidently saw nothing of interest, burst into a fit of coughing, staggered from the room, and was seen no more. The coughing ceased as soon as he left the lecture room.

Walter Rudin of the University of Wisconsin was a great teacher. He had many Ph.D. students, and each of his graduate courses was an event. Walter always liked to encourage questions from his audience. One day, at the beginning of the semester in his real analysis class, he said to the class, "Look. Don't just sit there. Ask questions. All questions are good. I like smart questions, I like stupid questions, I like all questions. Don't be timid. I never judge people by the questions they ask. I love them all. Just go ahead." Walter went on with his lecture, and after a while one young fellow screwed up his courage and raised his hand. Rudin recognized him, and the student posed his question. Walter gave a great smile, turned to the class, and said, "See. That's what I mean. That was a *really stupid question*."

Every college math major hears the story of Évariste Galois (1811-1832) being wounded in a duel and then perishing at the age of twenty. The story is that, knowing that he would likely be bested in the duel, he recorded his mathematical ideas the night before. Less well known is that when Galois was taken to the hospital with his fatal wounds, his brother waited there weeping at his bedside. Galois said, "Don't cry. I need all my courage to die at twenty."

D. V. Widder (1898–1990, Professor at Harvard) was invited by J. D. Tamarkin (Professor at Brown) to dinner and a symphony in Providence. Evidently Widder drove too quickly and was arrested in North Attleboro. He

Walter and Mary Ellen Rudin

was indicted on the spot and put in jail. Widder phoned Tamarkin, who borrowed bail money from the Brown University Treasurer, drove to the jail, and got Widder released. Meanwhile, Professor Widder wrote his lecture for the next day while he sat waiting in the hoosegow. He always supposed that his standing with the students improved when they heard the story.

Tamarkin and Widder took Fourier analyst L. Fejér (1880–1959) on a tour of the House of Seven Gables in Salem, Massachusetts. As they toured the various parapets and secret passages, Fejér always insisted on being last in line. They assumed that this was some quaint, old world politeness, or perhaps that he wanted to do some independent exploring on his own. When they returned to the car, they discovered that the real reason was that Fejér had sat on his sack lunch and got cherry juice all over the seat of his pants.

Widder tells that when World War I came he had to leave his studies at Harvard. He became a "civilian computer" in the range firing section at

Aberdeen Proving Grounds. Oswald Veblen was in charge, and Widder was bunked in a barracks with Norbert Wiener and Philip Franklin (1898–1965). He learned a lot from them, but they often inhibited his sleep by talking math far into the night. On one occasion Widder hid the light bulb, hoping to induce earlier quiet.

It is said that Einstein was a late talker. His first words, spoken well into his second year, were, "The soup is too hot." Einstein's much relieved parents asked why he had never said a word up until that moment. "Because," said Albert Einstein, "up to now everything was in order."

At the conference for Michael Atiyah's sixtieth birthday, Nigel Hitchin (an Atiyah student) related some memories of his association with Atiyah. From his undergraduate days, he pulled out his notebook from a course on manifolds. He declared that one could see Atiyah's style even in this undergraduate class. To illustrate the point, he began flipping through the notebook—reading aloud as he went. He found descriptions of and comments on the sphere, the plane, the line, the torus, homogeneous spaces, and so forth. However, Hitchin explained that he had gone through the entire notebook and never found an actual definition of "manifold."

When I was a graduate student, I was once having a conversation with Jeremy Stein (son of my thesis advisor E. M. Stein, and now an endowed chair Professor of Finance at MIT). He asked me for some help with a math problem (he was in the eighth grade at the time). I did my best to give him a cogent answer, as I did not want any negative reports to filter back to his father. Jeremy thought about what I told him and then he said, "You know, that was terrific. I asked you a question and you gave me an answer. Whenever I ask my Dad a math question he gives me a big song and dance. He won't just tell me what I want to know." Somewhat nervously, I said, "You know, your father is one of the greatest mathematicians in the world. You should try to appreciate what he tells you." Without hesitation, Jeremy said, "I don't think he's one of the greatest in eighth grade math."

In the 1960's, the Mathematical Association of America made a series of delightful one-hour films in each of which a great mathematician gave a lecture, for a general mathematical audience, about one of his achievements. One of these films starred Besicovitch, and he explained his solution of the Kakeya needle problem (S. Kakeya, 1886–1947). [For the record, the problem is this: Find the planar region of least area with the property that a segment of unit length lying in the region can be moved through all direction angles θ, $0 \leq \theta \leq 2\pi$, within the region. Besicovitch's surprising answer was that, for any $\varepsilon > 0$, there is a such a region with area less than ε.] Besicovitch was a natty dresser under any circumstances, and he wore to this lecture an attractive beige suit. However the lights were hot and, after a while, he removed his jacket to reveal bright red suspenders. The producers were most surprised (this was forty years ago, and nobody but firemen wore red suspenders), but the filming continued and the suspenders can be seen today.

At one point during the filming of Besicovitch, the aged professor had to blow his nose. He drew a large white handkerchief from his pocket and proceeded to do so—loudly. Later, when Besicovitch viewed the finished film, he objected to the noseblowing scene as undignified; he wanted it removed. The producers were able to replace the offending video segment, but it was decided that the sound should remain. As a result, if you view the film today, there comes a point in the action where the camera abruptly leaves Besicovitch and focuses on the side of the room—and you can *hear* Besicovitch blow his nose.

Robert Melter tells me that he once went to Harvard to listen to a seminar. He was standing in the hall, waiting for the lecture to begin, when Wiener started walking towards him. He was moving very slowly and it must have taken him two minutes to cross twenty feet. He got to within eight inches of Melter's nose and then grunted. It was a case of mistaken identity.

Wiener used to go three or four afternoons per week to the movie theater in Harvard Square and sit through the double feature. He claimed that he got many of his best ideas sitting in the back of the nearly empty theater.

Wiener's classes had to be held in large classrooms or small lecture halls. Although only a small number of students enrolled in Wiener's classes, there were many guests and auditors. Wiener was a formidable legend around campus and across Cambridge. He had had many outside offers, and rumor had it that his salary was \$100,000—even more than the Institute President. Wiener was the only Jew on the MIT faculty.

Joe Kohn, my teacher, tells of taking a class in Fourier analysis from Norbert Wiener. One day Wiener began his lecture by saying, "Today we will learn some applications of Fourier analysis to number theory. The basic unit of number theory is the prime number. A prime number is a positive integer that has no divisors except for 1 and itself. Here are some examples of primes..." And Wiener proceeded to write out, by hand, the first 200 primes. This took quite a while. Then he said, "The first result is this. If $\{p_1, p_2, \ldots\}$ are the primes, then

$$\sum_j \frac{1}{p_j} = \infty.$$

This is obvious by inspection."

One day Joe Kohn, a sophomore, was reading the newspaper alone in a student lounge at MIT. Wiener entered the room and began to walk around. Presently he approached the somewhat intimidated Joe Kohn and said, "Young man! Do you play the game of chess?" Kohn allowed that he did, and Wiener exclaimed, "Then let us play a game of chess!" Kohn relates that he was scared to death, and sure that Wiener would make a fool of him. About fifteen moves into the game, Wiener moved his queen so that Kohn's pawn could take it. Kohn was bewildered. Surely Norbert Wiener was making a devilishly clever sacrifice and would then get a quick checkmate. Kohn spent twenty minutes feverishly peering at the board and trying to determine what Wiener was up to. Finally he said, "Professor Wiener, I'm stumped. Why are you sacrificing your queen?" Wiener's eyes grew wide and he said, "Oh my God. That's a mistake! Can I take that move back?" Of course Kohn let Wiener retract his move. It became rapidly clear that Wiener was actually quite an inept chess player, and he quickly lost.

Norbert Wiener

Although Wiener was eccentric in many ways, he is also remembered as being extraordinarily kind. A graduate student, known only as "Moe," recalls bumming a ride from Wiener to get a free trip to New York City. Wiener was happy to offer the ride. They stopped in New Haven for dinner, and Wiener insisted on treating the impoverished young man to a nice steak dinner. He then insisted that they attend a movie together (they were only 100 miles from their destination). This they did, as a result of which they reached their goal at 2:00 a.m.

In the same vein, Norman Levinson tells of Wiener's great kindness in befriending him when Levinson was still a student. After Levinson showed some aptitude for mathematics, Wiener gave him the Paley-Wiener manuscript to read. Levinson found an error and proved a lemma to fix it. Wiener himself got out a typewriter and typed up the lemma for submission to a journal. Later, Wiener paid several visits to Levinson's family to introduce himself and to assure them of Levinson's future in mathematics. Levinson's parents were poor, ignorant immigrants. Jobs in academia were few, and Wiener showed a great kindness and understanding in helping these people to appreciate their son's talents.

It is well known that Wiener collaborated with R. E. A. C. Paley (1907–1933). The Paley-Wiener theorem has been extremely influential in harmonic analysis, and the book of Paley and Wiener is a masterpiece. Paley was a wild and reckless individual, and he died young. He used to torment Wiener in the following way. They would be working on a problem together, and would sometimes be stuck for hours. After a while, Paley would figure out how to solve the problem, but he would not tell Wiener. Instead he would say, "Norbert, this looks hopeless. I'm going to New York City for the weekend, where I will drink and chase women. See ya." This would plunge Wiener into the blackest despair. But on Monday Paley would return with the problem solved, and their collaboration would continue as usual.

Fred Almgren (one of my graduate school teachers) has inspired many stories. Surely his directness and disarming candor were a part of what makes it fun to tell Almgren tales. For example, when I first showed up as a graduate student at Princeton I thought I was hell on wheels. Almgren was my "part zero advisor" (meaning that he was to be my advisor before I had a formal thesis advisor). He sat me down on a sofa in his office, beneath a colorful photo of the fighter jet that he had piloted for the Navy. Almgren was, characteristically, dressed in a grey suit. I—straight from the University of California at Santa Cruz—was dressed like a rag picker. He said to me, "I don't know what kind of education you got at that Santa Cruz place. But let me tell you that we had serious doubts about admitting you to Princeton. Let me go on to say that at Princeton we do mathematics—no politics, no demonstrating, no fooling around. Just math." For me this was rather an abrupt introduction to Princeton mathematical life. I frankly did not know what to make of Almgren. Over time we became great friends, and I believe that he is the finest man that I have ever known.

Fred Almgren's son Robert is also a mathematician. He says, "A lot of fathers give life advice to their sons. My father used to tell me: 'Son, when you're in doubt, when you don't know which way to turn or what to do, I want you to remember two things. First, draw a picture. Second, integrate by parts.'" On another occasion, Fred said to his son Robert, "I don't put any pressure on my son to follow me in my work. He can do anything he wants to. He can be an algebraist, a topologist, a geometer, ..."

Traditionally, in Sweden, there have been a few surnames that have been widely used. "Jenson" and "Swenson" are among these. In fact there were, at one time, so very many Jensons and Swensons that the government had a difficult time keeping track of people. So it passed a law encouraging people to change their names. The Swedes are a practical people, and they adapted readily to this new paradigm. Typically, on reaching the age of majority, or on getting married, people would sit down and choose a new name. In some cases the entire group of children in a given family would sit down and choose a new name. The story is that this latter is in fact what the siblings who are now known as Hörmander did. The Fields Medalist Lars Hörmander was not born with that name. Likewise his famous student Christer Kiselman (1939–) was actually born with the name "Swenson." He chose the name "Kiselman," which means "silicon man."

Someone once hurried to tell Carl Friedrich Gauss that his wife was dying. He found the celebrated mathematician deep in thought, no doubt pre-occupied with a theorem. The courier blurted out the dire news. "Tell her to wait a few minutes until I have finished," was Gauss's reply.

In India there are a great many people—many millions in fact—who are named "Gupta." One year a clerical error was made in the mathematics graduate office at Washington University in St. Louis (my institution) and the wrong Gupta was admitted to the mathematics Ph.D. program. How was the error discovered? Well, the letters of acceptance went out in a given week, and several weeks later the responses began to come in. The Gupta in question responded

> Joy reigns supreme in my village. Nobody in this district has ever been to school before. We have sacrificed a cow.

Somehow we knew we had the wrong Gupta. It was difficult to do, but somebody had to send a letter of apology to Gupta and explain that in fact he or she did not qualify for our Ph.D. program. There was nothing that could be done for the cow.

Many years later, I related this story to a math department visitor who had recently arrived from New Delhi. He listened with a grave face, and did not find the tale at all amusing. At the end he nodded slowly and said, "You know, in India there are entire districts where everyone is named Gupta." I replied, "That's amazing. How do you tell people apart?" Then he fixed me with a particularly dour expression and said, "Why is there any need to tell people apart?"

Lars Ahlfors was arguably the most important complex function theorist of the twentieth century. Loved by all, he was also an ardent drinker. One day he walked from his home in the posh Back Bay area of Boston to buy a bottle of scotch. On his return, walking up his front steps, he was assaulted by a young man with a knife. Never one to hesitate, Ahlfors bashed the fellow on the head with the bottle of scotch and fled into the house. He telephoned the police and the man was arrested. Next day, Ahlfors took great pride in telling of his exploits. But it was clear that what he truly regretted was the loss of the scotch.

Once, at a conference, Ahlfors stayed up late drinking with friends. He was drinking beer. At one point late in the evening his wife walked up, handed him a bottle of scotch, and said, "Here, this is quicker."

A conference was held in Storrs, Connecticut to honor Lars Ahlfors on his 70th birthday. After the banquet, Ahlfors rose to make a few remarks. He began by saying, "Retirement is great. I can no longer perish so I don't have to publish." Ahlfors cherished the fact that he could devote all his time to studying the ideas that interested him most, without expending the effort to write up his results. He seemed to think of this as a mathematician's paradise. Ahlfors met a friend of mine at the beginning of the banquet evening, and again at the end. At the second encounter he had no memory of the first meeting, and my friend protested. Ahlfors shrugged and said, "Ah, but there are so many of you and just one of me."

I was once lunching with an NSF program officer, and he was lamenting the reluctance of famous mathematicians to put any effort into writing up their NSF proposals. He went on to say, "Let me tell you about the proposal that I just got from Ahlfors." I thought, "Well, this is Ahlfors. It must be a doozy." In fact the proposal said, "I will continue to study the work of Thurston." That was it—no explanation, no references, no nothing.

A different NSF officer once described a typical NSF proposal from Harvard or Princeton or MIT as saying, "Here's my address. Where's my check?"

Lars Ahlfors and Arne Beurling were great, lifelong friends. Ahlfors described the difference in their personalities as follows:

> Our personalities were as different as could be. I was the traditional stodgy, mostly silent Finn, chronologically two years younger, but in reality so immature that I was like a child compared to Arne ...
>
> Typically, after obtaining my first academic degree, I was very happy that my father made it possible for me to go with my teacher Rolf Nevanlinna to Zürich, where he was substituting for Hermann Weyl at ETH...
>
> Meanwhile, Arne's father took his son to Panama to hunt alligators, a rather different approach to parental upbringing. Arne loved it, while I, in his place, would have tried to get as far away as possible from all guns and alligators.

Fields Medalist Alexander Grothendieck (1928–) had many eccentricities. His student Robin Hartshorne once asked Grothendieck a mathematical question. The famous man said, "It's either obvious or it's false."

One day Grothendieck lectured for 50 minutes, making reduction after reduction to prove a theorem. Finally he looked up and said, "Who am I kidding? This is false."

In 1988, Grothendieck was awarded the prestigious Crafoord Prize (an adjunct to the Nobel Prize that admits of prizes for mathematics). He turned it down, deriding scientific nepotism, dishonesty, and overwhelming politics. He also pointed out that he did not need the money. Today, Grothendieck is retired and has moved to some remote part of the world. Nobody knows where he is.

In the 1950's, there was a math department Christmas party at the University of Chicago. Many distinguished mathematicians were present, including André Weil, S. S. Chern, and Irving Segal (1918–1998). For amusement, the gathered company endeavored to assemble a list of the ten greatest living mathematicians, present company excluded. Weil, however, insisted on being included in the consideration.

The company then turned to the assemblage of a list of the ten greatest mathematicians of all time. Weil again insisted on being included.

Weil later moved to the Institute for Advanced Study in Princeton. As was the custom, Weil often attended tea at the University. Graduate student Steven Weintraub one day went about the room asking various famous mathematicians who was the greatest mathematician of the twentieth century. When he asked Weil, the answer (without hesitation) was "Carl Ludwig Siegel (1896–1981)." When Weintraub then asked who was the *second greatest* mathematician of the twentieth century, Weil just smiled and proceeded to polish his fingernails on his lapel.

George Dantzig (1914–) is a Professor at Stanford University, and credited with inventing the subject of linear programming. Many years ago, Dantzig was a student and one day he arrived late for a math exam. The professor was the Polish statistician Jerzy Neyman (1894–1981). There were three problems written on the blackboard. Dantzig solved the first two problems easily but had to struggle with the third; he barely solved it before the end of the test period. He put his paper on the professor's desk and left the room. Little did he know that the three problems on the board were famous unsolved problems that the professor had recorded for cultural purposes. The actual exam was handed out on paper.

Some time later, Neyman read Dantzig's paper and was dumbfounded. He contacted Dantzig, praising him for what he had achieved. Later, Dantzig was told to write an introduction to his test paper and it became Dantzig's Ph.D. thesis.

A similar story has been told about the topologist John Milnor (1931–) at Princeton, with Ralph Fox as the Professor. A few years ago, Milnor was asked whether the story was really true. The gist of his answer was, "No, but it *could* have happened."

George Pólya tells this story of his former student John von Neumann: "He is the only student of mine I was ever intimidated by. He was so quick. There was a seminar for advanced students in Zürich that I was teaching and von Neumann was in the class. I came to a certain theorem and I said it is not proved and it might be difficult. Von Neumann didn't say anything

Jerzy Neyman

but after five minutes he raised his hand. When I called on him he went to the board and proceeded to write down the proof. After that, I was afraid of von Neumann."

Michigan State University often gets rather brilliant undergraduates. In the early 1960's, a couple of math majors—one of them named Peter Rheinstein—developed the habit of attending faculty seminars and colloquia. One day the speaker was an avatar of what one might call "bad algebra." He wrote twelve axioms on the board for some new gadget that he called a "hemi-semi-demi something-or-other" and proceeded to prove propositions and theorems about it. After about twenty minutes, Rheinstein strode to the blackboard and wrote down a three-line proof that hemi-semi-demi something-or-others do not exist. The assembled faculty looked at each other in an embarrassed fashion and then slowly (and silently) filed out of the room.

Stefan Bergman was a distinguished, if eccentric, complex analyst whose work has been extremely influential. He had command of many languages, and made sure that everyone knew it. When Waclaw Sierpiński (1882–1969) spoke at Stanford about Hypothèse *H*, he spoke in fractured French. In fact it was so fractured that it was quite easy for most Americans to understand—just by transliteration. But Bergman insisted on translating Sierpiński's remarks into English. Unfortunately, Bergman's translation was much less clear.

Bergman also provided a translation of a three-hour presentation by a visiting Russian. Afterward, Bergman declaimed that, "I speak seven languages, English the bestest."

Bergman was a native of Poland. He was once conversing—in Polish—with his fellow Pole Antoni Zygmund. After a time, Zygmund said, "Please, let us speak English. It is more comfortable for me."

Bergman's wife Edy was quite devoted to him, but life with Stefan was sometimes trying. When they first got married, Bergman had just completed a difficult job search. In the days immediately following World War II, jobs were scarce; and Bergman wanted a position with no teaching. After a long period of disappointment, Max Schiffer got Bergman a position at Stanford; so the mood was high at the Bergmans' wedding reception. The celebration took place in New York City, and Bergman was delighted that one of the guests was a mathematician from New York University with whom he had many mutual interests. They got involved in a passionate mathematical discussion and after a while Bergman announced to the guests that he would be back in a few hours: he had to go to NYU to discuss mathematics. On hearing this, Schiffer turned to Bergman and said, "I got you your job at Stanford; if you leave this reception I will take it away." Bergman stayed.

There is considerable evidence that Bergman thought about mathematics constantly. Once he phoned one of his graduate students, at the student's home number, at 2:00 a.m. and said, "Are you in the library? I want you to look something up for me."

Bergman was a prolific writer. Of course he worked in the days before word processors and before TEX. His method of writing was this: First, he would write a manuscript in longhand and give it to the secretary. When she had it typed up he would begin revising, stapling strips of paper over the portions that he wished to change. Strips would be stapled over strips, and then again and again, until parts of the manuscript were so thick that the stapler could no longer penetrate. Then the manuscript would be returned to the secretary for a retype and the whole cycle would begin again. Sometimes it would repeat *ad infinitum*. Bergman once told a student that, "a mathematician's most important tool is his stapler."

The noted Danish mathematician Harald Bohr had typically cryptic European handwriting. Many words just looked like jagged lines, with no

Harald Bohr

discernible letters. One day he was giving a lecture and said, "And so this point is a maximum." The word "maximum" was written as a jagged line, with no adornments. Someone raised his hand and said, "Professor Bohr, I think it's a minimum." "Oh yes," said Bohr, adding two dots.

In 1962, Swedish mathematician Lars Hörmander won the Fields Medal for his brilliant work in partial differential equations. The Swedish academic system is different from the American one in that there are a fixed number of Professorships in mathematics—about twenty—in the entire country and a new person cannot become a Professor unless an existing Professorship is vacated through retirement or resignation or death. At the time, there was no Professorship available for Hörmander so he accepted a job at the Institute for Advanced Study in Princeton. It wasn't long before the Swedes perceived this situation to be a national tragedy. A member of the Swedish Parliament argued vehemently in session that one of the greatest living mathematicians was a Swede and his home country had no job to offer him. So a new Professorial position, that came affectionately to be known as the "Hörmander Professorship," was created for Lars Hörmander at the University of Lund. Hörmander accepted the job, and stayed there for the remainder of his career.

Stanislaw Ulam's Aunt Caro is buried in an elaborate and impressive mausoleum in Monte Carlo. She was directly related to the famous Rabbi Loew of sixteenth-century Prague, who, the legend says, made the Golem—the earthen giant who was protector of the Jews. Norbert Wiener, knowing that Ulam was involved with the creation of the hydrogen bomb, and learning of his very special family history, commented, "It is still in the family."

Ulam's friend John von Neumann was quite wealthy, and lived a plutocratic life. He liked to say, "It is not enough to be rich, one must also have money in Switzerland."

In January of 1913, Hardy received an unsolicited letter from Madras. What was written is of such great historical import that we record it here:

I beg to introduce myself to you as a clerk in the Accounts Department of the Port Trust Office at Madras on a salary of only [20 pounds] per annum. I am now about 23 years of age. I have had no university education, but I have undergone the ordinary school course. After leaving school I have been employing the spare time at my disposal to work at Mathematics. I have not trodden through the conventional regular course which is followed in a university course, but I am striking out a new path for myself. I have made a special investigation of divergent series in general and the results I get are termed by the local mathematicians as 'startling'...

Very recently I came across a tract published by you styled *Orders of Infinity* in page 36 of which I find a statement that no definite expression has been as yet found for the no of prime nos [sic] less than any given number. I have found an expression which very nearly approximates to the real result, the error being negligible. I would request you to go through the enclosed papers. Being poor, if you are convinced that there is anything of value I would like to have my theorems published. I have not given the actual investigations nor the expressions that I get but I have indicated to the lines on which I proceed. Being inexperienced I would very highly value any advice you give me. Requesting to be excused for the trouble I give you.

—Srinivasa Ramanujan

The correspondent, S. Ramanujan (1887–1920), was completely unknown to Hardy. When Ramanujan was sixteen he came by chance on a copy of Carr's *Synopsis of Mathematics*, and this proved for him a seminal inspiration. This book provided Ramanujan with a complete account of the formal side of integral calculus, including Parseval's formula, Fourier inversion, and related material. Ultimately, Ramanujan sent to Hardy a long list of very difficult integrals with techniques for evaluating them. Hardy, a master of this type of problem, was able to do most but not all of them. Hardy and Littlewood quickly concluded that Ramanujan was an exceptional talent, and they brought him to Cambridge. Hardy was soon convinced that, measured by natural ability, Ramanujan was in the class of Euler and Gauss. Hardy and Ramanujan wrote many splendid papers together in number theory and related subjects. Littlewood never collaborated with Ramanujan. In May, 1917, Ramanujan fell ill. He returned to India in 1919 and died in 1920.

It is not generally well known that Ramanujan actually wrote from Madras to a number of distinguished mathematicians (besides Hardy). These included the applied analyst M. J. M. Hill, the geometer H. F. Baker, and the analyst E. W. Hobson (1856–1933). But only Hardy thought it worthwhile to reply in full to Ramanujan's letter.

Ramanujan's widow was much younger than he, and survived him by a great many years. Mathematician George Andrews went to India and met with her around 1980.

It was once said of Ramanujan that each of the positive integers was one of his personal friends. There is disagreement over whether it was Hardy or Littlewood who said it. Littlewood claims that he delivered the line to Hardy, who received it with a grim silence and a poker face. Littlewood then teased Hardy about it, and the reply was, "Well, what is one to do, is one always to be saying 'damned good'?" Littlewood's rejoinder was "Yes."

John E. Littlewood was born in 1885. From about 1930 on, his happiness was considerably marred by an obscure nervous malady which afflicted him for about thirty years. It was not apparent to most people that he was at all ill. His lectures were still of the highest quality, he had a number of Ph.D. students, he performed many difficult and demanding tasks at the university, and he continued to be creative at the highest level. He was awarded the Sylvester Medal in 1943 and the Copley Medal in 1958, attesting to the fine quality of his work. Even so, Littlewood was adamant in asserting that he was functioning at half his capacity for about thirty years; he frequently spent hours in cinemas just to while away the tedious hours.

In the year 1960, Littlewood suffered a severe attack of influenza. His physician, Edward Bevan, learned thereby of Littlewood's depression and referred him to the neurologist Beresford Davies. Davies found a combination of drugs that cured Littlewood's ills. This changed his life, and he accepted several invitations to visit the United States for the first time (at the age of 75).

Littlewood was a creature of many habits. One of these is that, each afternoon, he would drink a large glass of vodka diluted with water. This is a habit that he acquired from his colleague Besicovitch. Littlewood listened only to Bach, Beethoven, or Mozart; he considered life to be too short to waste on other composers. Littlewood always said that one could not do mathematics for several hours after lunch, because the blood cannot be in two places at once.

My colleague Al Baernstein, a noted complex analyst himself, once had the pleasure of meeting J. E. Littlewood. He tells of going to Littlewood's hotel room and of having to clamber over numerous vodka and other liquor bottles in order to shake the great man's hand.

On summer vacations with Ann Streatfeild in Treen, Littlewood began the day with coffee in bed. He would then go to work on the sun porch, where he had a broken-down chair, a log on which to put his feet, and stones to weigh down his papers. He then went for a swim, timing himself exactly. He went for another swim in the afternoon. If the weather was not good for swimming, then he would go for a walk of pre-specified length. After tea he would play Patience, then fetch beer to drink with his evening meal. The evening was spent playing card games.

Littlewood recounts that when G. D. Birkhoff (1884–1944) met you he would say, "Sir, you are the greatest mathematician in the world." The expected reply was, "With one exception."

J. J. Sylvester sent a paper to the London Mathematical Society for publication. True to form, he included a cover letter asserting that this was the most important result in the subject for 20 years. The Secretary replied that he agreed entirely with Sylvester's assessment, but that Sylvester had actually published the result in the *Journal of the London Mathematical Society* five years earlier.

Landau was very hard on his Privat Dozents. Once one of them was recuperating in the hospital, and it is said that Landau climbed a ladder and pushed work for him to do through the window. On another occasion, one

Edmund Landau

of the Privat Dozents was going on his honeymoon. Landau wanted him to take the proof sheets of his (Landau's) three-volume book on number theory along for proofreading. The new bride was quite distressed, but Landau would hear none of it. Finally Landau's wife intervened; in the end it was agreed that the Privat Dozent would take along only one volume.

Pólya attended a lecture of Einstein, where he heard, "This has been done elegantly by Minkowski; but chalk is cheaper than grey matter, and we will do it as it comes."

G. H. Hardy claimed that he "thought on paper," with his pen. He wrote everything out (in his beautiful hand), scrapping and re-copying whenever the page got into a mess. In fact Hardy was known to copy out printed proofs from a journal, or even printed proofs from one of his joint works with Littlewood. This harkens back to classical times, when students trained by copying the works of the great scholars.

Littlewood, by contrast, claimed that when he was thinking about a problem all his thoughts would go onto a single page—all over the place with odd equations, diagrams, rings. However appalling the mess, he felt that to scrap that page would somehow break threads in the unconscious.

In 1912 an insane Fellow of Cambridge named P. J. Pearce was thought to have escaped from his asylum and likely to come to a College meeting at the university with a revolver. The doorkeepers chosen for toughness were H. F. Stewart and J. E. Littlewood with A. V. Hill as the runner up.

In 1952, J. E. Littlewood dreamt that he was at a party. Catching sight of himself in a mirror, he found that he was wearing a halo. His reaction was that he hardly should have thought he merited such an honor, but who was he to question holy authorities?

Littlewood was once asked who were the greatest intellects, who had had the biggest influence on his thinking. Of course he mentioned Archimedes and Newton. But, he allowed, as a matter of fact the truly greatest man was the person who had taught him algebra in school.

Littlewood and his group had booked five reserved third class seats on the train. On arrival, they found the seats occupied. Through some error, the seats had been booked twice. They protested to an official, with little effect. Then one member of the party went to the headquarters of the rail company and remonstrated that, "Professor Littlewood is very much annoyed." Presently a young man in plus fours appeared, saying, "Professor Littlewood, it is one of the proudest moments of my life to meet you at last, I was a pupil of yours at Manchester." The group was put in a reserved First Class compartment, no-one else admitted, and lunch brought to them by a special service.

Gian-Carlo Rota recalls Alonzo Church (1903–1995) from his undergraduate days at Princeton. Church looked like a cross between a panda and a large owl. He spoke softly in complete paragraphs, evenly and slowly enunciated. He never made casual remarks, for they could not be part of formal logic. He would never say, "It is raining." Taken in isolation, such a statement would make no sense for Church. Instead he would say, "I must postpone my departure for Nassau Street, inasmuch as it is raining, a fact which I can verify by looking out the window."

Church's course in mathematical logic was a fixture of mathematical life at Princeton. Every lecture began with a 10-minute ceremony of erasing the blackboard until it was absolutely spotless. Sometimes the students would try to save him the effort by erasing the blackboard beforehand. This did not work. The ritual, performed by Church himself, could not be dispensed with. Often it required soap, water, and a brush; it was followed by another ten minutes of total silence while the blackboard was drying. Some thought that Church needed the time to prepare his lecture in his mind. This seems unlikely, as his lecture was an accurate rendition of the typescript that resided in the Fine Hall library.

In Church's time, the Princeton Mathematics Department had mixed views of mathematical logic. Church's course was never heavily subscribed. Two minutes before the end of Church's lecture (which met in the largest classroom in Fine Hall), Lefschetz would begin to peek through the door. Lefschetz would glare at the students and at the spotless text on the blackboard; sometimes he shook his head as if to indicate that he considered the enterprise to be a lost cause. The following class was taught by Kunihiko Kodaira, whose work in geometry was revered at the time.

At the same time John Kemeny (1926–1992, a student of Church) was a departing instructor in the mathematics department, being eased into philosophy by Lefschetz. Kemeny's course in the philosophy of science concentrated on basic reasoning. Kemeny was not afraid to appear pedestrian, trivial, or stupid. His main concern was to respect the facts, to make distinctions even when they went against our prejudices, and to avoid oversimplifications. "There is no reason why a great mathematician should not also be a great bigot," said Kemeny. "Look at your teachers in Fine Hall, at how they treat one of the greatest living mathematicians, Alonzo Church."

The algebraist Emil Artin (1898–1962) cut a dashing figure; with his long winter coat cinched around his waist, light blue eyes, and gaunt face he reminded one of nothing so much as a Wehrmacht officer. He gave rigorous, axiomatic lectures with no motivation and few examples. He was often tough and rude to his students, creating embarrassing public scenes with his fits of temper. He would throw a piece of chalk or a coin at a student who asked a silly question. For foolish statements he delivered shouted diatribes. One famous example occurred when his Ph.D. student Serge Lang (1927–) professed that Pólya and Szegő's *Aufgaben und Lehrsätze* was bad for mathematical education. In fact Artin loved special functions and explicit computations (though he avoided them religiously in class). He cherished that book.

As a young man, in his closest circle of friends, Emil Artin had the nickname "Ma." He always preferred this name to his given name of Emil. The name is short for "Mathematics." To his young male friends, Artin was the embodiment of mathematics.

Gabor Szegő

William Feller was one of the fathers of modern probability theory. His seminal text on the subject is still a standard reference. He has been a most influential teacher.

His lectures were loud and entertaining. He wrote very large on the blackboard in a florid script, with lots of whirls. Sometimes only one formula appeared on the blackboard during the entire hour. The rest was what we might call handwaving. His "proofs" were sketchy and vague, but they were convincing. He always got the main idea across.

If someone interrupted Feller's lecture by pointing out a mistake, he would become quite incensed. He became red in the face and raised his voice, often to a full shout. Sometimes he would ask the interrupter to leave the classroom. Mark Kac coined the term "proof by intimidation" to describe Feller's style. Feller made the students feel that they were being let in on some wondrous secret, and there was a sense as he left the room at the end of the hour that some magic was vanishing with him.

Feller often interrupted his lectures with little tirades about ancillary topics. He gave these diatribes titles, including "Gandhi was a phony," "Velikovsky is not as wrong as you think," "Statisticians do not know their business," "ESP is a sinister plot against civilization," "The smoking and health report is all wrong," and "The Babylonians knew Fourier analysis."

In point of fact, Feller and Velikovsky (1895–1958) were neighbors in Princeton. They first met one day when Feller was working in his garden, pruning some bushes. Velikovsky came rushing out of his house screaming, "Stop! You are killing your father!" They soon became close friends.

Solomon Lefschetz's most universally recognized trait was his rudeness. He was rude to everyone, and sometimes obnoxious to boot. In 1957 he met Oskar Zariski, the man who formulated the ideas that led to Hironaka's celebrated resolution of singularities theorem. After exchanging with Zariski warm and loud Jewish greetings (in Russian), Lefschetz proclaimed vigorously his skepticism of the possibility of proving resolution of singularities for all algebraic varieties. "Ninety percent proved is zero percent proved!" he replied to Zariski's protestations.

When news first came to Princeton of Hodge's (W. V. D. Hodge, 1903–1975) work in England on harmonic integrals and their relation to homology, Lefschetz dismissed it as the work of a crackpot. In fact his statement was such a gaffe, and since proved completely wrong, that Lefschetz toned down his proclamations a bit after that.

When Spencer and Kodaira were still Associate Professors at Princeton, they proudly explained to Lefschetz a new proof they had found of one of Lefschetz's deeper theorems. "Don't come to me with your pretty proofs. We don't bother with that baby stuff around here," was Lefschetz's irate reply. Nevertheless, Lefschetz showed considerable regard for the two young mathematicians after that.

Lefschetz told with some relish the story of one of E. H. Moore's (1862–1932) visits to Princeton. Moore began a lecture by saying, "Let *a* be a point and let *b* be a point." Lefschetz shouted, "But why don't you just say, "Let *a* and *b* be points?" Moore replied, "Because *a* may equal *b*." Lefschetz got up and left the room.

Lefschetz's lectures were well known to be intuitive, vague, and often incoherent. One student remembers a course on Riemann surfaces. Lefschetz began with a string of statements in rapid succession, with nothing written on the blackboard: "Well, a Riemann surface is a certain kind of Hausdorff space. You know what a Hausdorff space is, don't you? It is also compact, OK? I guess it is also a manifold. Surely you know what a manifold is. Now let me tell you one nontrivial theorem: the Riemann–Roch theorem." And so forth, until all but the most faithful students dropped the course.

Gian-Carlo Rota remembers attending one of Lefschetz's classes. Of course Lefschetz wrote with a piece of chalk jammed into one of his artificial hands. He wrote enormous letters on the blackboard, and it was difficult to follow anything he was saying. Rota became discouraged, and at one point asked a senior mathematician who had been attending whether he understood what was going on. The senior man gave a vague and evasive answer. Then Rota understood.

Lefschetz was ultimately forced to retire from his chairmanship at Princeton because of his age. He developed a love for Mexico and decided to help develop Mexican mathematics. Once, in a Mexican train station, he singled out a charro dressed in full regalia, with a pair of pistols and a cartridge belt across his chest. He started ridiculing the charro, adding some deliberate slurs in his excellent Spanish. Of course charros are well known for reacting quickly and decisively to insults. Lefschetz's companions feared for his safety. Indeed, the charro eventually stood up and reached for his pistols. Lefschetz looked him right in the eye and did not budge. There were a few moments of tense silence. Then the charro exclaimed, "Gringo loco!" and stalked away. After Lefschetz left Mexico, he was awarded the Order of the Aztec Eagle.

Lefschetz had the highest mathematical standards, and he used them to help build the Princeton mathematics department into the pre-eminent department that it is today. When addressing an entering class of twelve graduate students, he said, "Since you have been carefully chosen among the most promising undergraduates in mathematics in the country, I expect that you will all receive your Ph.D.'s rather sooner than later. Maybe one or two of you will go on to become mathematicians."

The Directorship of the Institute for Advanced Study is not just an academic position. It is also a politically sensitive one. The person is a public figure, speaks for the American academic establishment, is involved in fund-raising, testifies before Congress, and hobnobs with bigwigs. So great care is exercised in choosing those who will occupy the office.

In the early 1980's, the Director of the Institute (Harry Woolf) was in the habit of walking his dogs at the end of the day. Andrew Neff, a Ph.D. student of E. M. Stein (my thesis advisor) at Princeton, was in the habit of going for a run each day in the same place at the same time. Unfortunately, the Director did not leash his dogs, and they had the nasty habit of assaulting the runner. Andy took umbrage at the situation, and spoke rather sternly to the Director about the matter, but the offense was continually repeated. He finally became quite exasperated and leveled a threat at the esteemed Director: if the dogs assaulted him again, he would take drastic action.

Next day, the Director walked his dogs and the student took his run and—guess what?—the dogs attacked the student. The student hauled off and slugged the Director of the Institute for Advanced Study, and a fist fight ensued. The student evidently won, but the worst part of the matter is that both parties were arrested. The whole thing got into *The Princeton Packet* (the local newspaper). The entire matter became a *cause célèbre*, hotly debated at math department and Institute for Advanced Study teas.

Everyone took a side. It is hard to say what effect these malodorous affairs had on the career of Harry Woolf, but in 1987 his stint as Director of IAS was ended.

During Halmos's Chicago days, there were many distinguished visitors. One was Raymond Jessen (Antoni Zygmund's collaborator). He was to lecture directly after Halmos, and Halmos erased the board in haste to make way for Jessen. Jessen helped, and the only thing he left was the phrase "almost periodic function" about two-thirds of the way down on the second board. At the appointed moment, Jessen began his lecture. About 35 minutes later, he reached the unerased part of the board, and—glory be!—"almost periodic function" was just the phrase he needed for his exposition.

Lars Gårding was a student of Marcel Riesz. One of his duties as Riesz's assistant was (during a time of rationing) to transfer his liquor allotment to Riesz.

Nachman Aronszajn (1907–1980) taught at the University of Kansas, in a town which was "dry"—no liquor. One of Aronszajn's great joys in life was good food and drink. He lamented that "young people of today" didn't know how to enjoy a good meal. If you visit the University of Kansas mathematics department today, you can see the "Aronszajn room," with beautiful paneling and oak-trimmed blackboards. If you are lucky, one of the *cognoscenti* will show you the secret compartment in the blackboard that was designed for hiding Aronszajn's bottles of liquor.

J. J. Sylvester (1814–1897) was educated in England. When he was still young, he accepted a professorship at the University of Virginia. One day a young member of the chivalry whose classroom recitation Sylvester had criticized became quite piqued with the esteemed scholar. He prepared an ambush and fell upon Sylvester with a heavy walking stick. Sylvester speared the student with a sword cane, which he just happened to have handy. The damage to the student was slight, but the professor found it advisable to leave his post and take the earliest possible ship to England. Sylvester took a position there at a military academy. He served long and well, but subsequently retired from his position and accepted a position at Johns Hopkins University when he was in his late fifties. He founded the *American Journal of Mathematics* the following year, and was certainly the leading light of American mathematics in his day.

Mark Kac tells that the attendance at Hugo Steinhaus's seminar was small. One day only Kac and one other student were present. This fact did not seem to bother Steinhaus. But, afterward, Kac asked him what was the minimal audience to which he would lecture. Steinhaus replied, "*Tres facit collegium*"—which means "Three makes a college." The next time only Kac was present; the other student had quit. As Steinhaus was beginning his lecture, Kac interrupted with, "What about '*Tres facit collegium*?'" Steinhaus answered that, "God is always present." It should be noted that Steinhaus was an outspoken atheist.

An eminent mathematician at Princeton, whose name will be kept confidential because of the statute of limitations, grew up in New York City and attended Stuyvesant High School—the estimable New York "magnet school" for students who are gifted in science. There he became a member of a secret society for which the rite of initiation was to steal a book from the Barnes & Noble bookstore. He stole a copy of Ince's (Edward Lindsay Ince, 1891–1941) celebrated book on ordinary differential equations. Today he is a famous analyst.

In like manner, there is a famous geometric analyst at Berkeley who was thrown out of high school for stabbing a fellow student. Seemed that the victim had put a broken-up lightbulb into the miscreant's luncheon sandwich. The stabbing was (perhaps) just revenge. After being expelled from school, the quite brilliant stabber was admitted to the University of Chicago. The rest is history.

William Chauvenet (1820–1870) founded the Naval Academy in Annapolis, Maryland. He then moved to Washington University in St. Louis, where he founded the Mathematics Department (it had been part of the Astronomy Department, as was the case at many American Universities of the time). After that he became Chancellor of the University. Part of

Chauvenet's scientific work at Washington University was to do all the calculations (by hand, of course) for the design of the Eads bridge. The bridge still stands today. William Chauvenet died at the age of fifty.

The Chauvenet Prize of the Mathematical Association of America is named after William Chauvenet. It is a prize for excellence in mathematical writing. Washington University has had more Chauvenet Prize winners than any other university: Guido Weiss, Ken Gross, and myself.

Ernst Eduard Kummer (1810–1893) was one of the important algebraists of the nineteenth century. But he was dreadful at arithmetic. Often in class he resorted to enlisting his students to help him with simple problems. Once he had to calculate 7 × 9. "Seven times nine," he began. "Seven times nine is, er, — ah — ah — seven times nine is ..." "Sixty-one," came a voice from the front row. Kummer wrote 61 on the blackboard. "Sir," interrupted another student, "it should be sixty-nine." "Come, come, gentlemen, it can't be both," Kummer exclaimed. "It must be one or the other."

Another version of the story is that Kummer reasoned it out. He said, "The product cannot be 61, because 61 is prime. It cannot be 65, because 65 is a multiple of 5. It cannot be 67 because it, too, is a prime; 69 is too big. Only 63 is left."

Henry Eyring (1901–1981) taught chemistry at the University of Utah. In his earlier days he had been a colleague of Einstein, and he told this story. One day they were on a walk together, and they came upon an unusual plant growing along a garden path. Dr. Eyring asked Dr. Einstein whether he knew what the plant was. Einstein had no idea, so together they consulted a gardener. The gardener told them, with a straight face, that the plant was green beans. After that, Eyring took great delight in telling people that he could prove that Einstein didn't know beans.

In a lecture, Paul Dirac (1902–1984) made a mistake in an equation that he recorded on the blackboard. A student raised his hand and said (rather timidly), "Professor Dirac, I do not understand equation (2)." Dirac gave no reaction, and simply continued to write on the blackboard. After a while, the student assumed that Dirac had not heard him. He repeated, rather more loudly, "Professor Dirac, I do not understand equation (2)." But there was still no reaction. Another student in the first row decided to intervene. He said, "Professor Dirac, that man is asking a question." Dirac paused and then he replied, "Oh, I thought he was making a statement."

Like most Germans, the Einsteins were subject to rather severe economies in post-World War I Germany. Mrs. Einstein even saved old letters and envelopes and scraps of paper so that Einstein could continue his calculations and the development of his ideas.

Years later, when Einstein was a founding member of the Institute for Advanced Study and a great celebrity in the United States, Mrs. Einstein was pressed into public service—doing public relations at important and influential science centers. Dutifully, she plodded through lab after lab filled with hundreds of thousands of dollars worth of gleaming equipment. As part of the ceremony, American scientists explained to her what all this over-priced finery was for.

On one of these excursions, Mrs. Einstein was ushered into a high-chambered observatory, and came right up to a huge, overwhelming scientific contraption. "Well, what's this one for?" she asked in some exasperation. "Mrs. Einstein, we use this equipment to probe the deepest secrets of the universe."

"Is *that* all?" snorted Mrs. Albert Einstein. "My husband did that on the back of old envelopes."

We all know that mathematicians are multi-faceted creatures. That guy down the hall who seems to think about Moufang loops all day may also be a concert pianist, or a closet horticulturist, or an expert in orienteering. Guido Weiss, the Eleanor Anheuser Chair Professor of Mathematics at Washington University, was drafted by the Chicago Bears as a young man. Gary Sherman of the Rose-Hulman Institute of Technology was drafted both by the Cleveland Browns and the Boston Patriots. For "religious reasons," Gary signed with the Browns. But a different religion ultimately beckoned, and he became a mathematician.

Frank Ryan was quarterback of the Dallas Cowboys. One fine year he led his team to a win at the Superbowl. It also turns out that Frank Ryan has a

J. P. Serre

Ph.D. in mathematics. A journalist once asked Ryan how his advanced degree aided him in the Superbowl victory. "Not one damned bit," was the demure reply.

Raoul Bott (1923–) describes J. P. Serre (1926–)—Fields Medalist, 1954—as one who has mathematical ideas so clear in his mind that he makes everything look like child's play. To make matters even more aggravating, Serre never seems to work. He is always seen playing ping pong or chess or reading the newspaper; he never appears to be laboring away. Mrs. Serre, by contrast, says that her husband works all the time. And Serre himself claims that he does all his best work in his sleep.

Further Reading

E. T. Bell, *Men of Mathematics*, Simon & Schuster, New York, 1965.

P. Duren, *A Century of Mathematics in America*, with the assistance of Richard Askey and Uta Merzbach, American Mathematical Society, Providence, 1988--1989.

H. Eves, *In Mathematical Circles*, Prindle, Weber, & Schmidt, Boston, 1969.

——, *Mathematical Circles Revisited*, Prindle, Weber, & Schmidt, Boston, 1971.

——, *Mathematical Circles Squared*, Prindle, Weber, & Schmidt, Boston, 1972.

——, *Mathematical Circles Adieu*, Prindle, Weber, & Schmidt, Boston, 1977.

——, *Return to Mathematical Circles*, Prindle, Weber, & Schmidt, Boston, 1988.

C. Fadiman, *The Mathematical Magpie*, Simon and Schuster, New York, 1962.

P. Halmos, *I Want to Be a Mathematician*, Springer-Verlag, New York, 1985.

G. H. Hardy, *A Mathematician's Apology*, Cambridge University Press, London, 1967.

M. Kac, *Enigmas of Chance*, Harper & Row, New York, 1985.

Karl-Franzens-Universität Graz Institut für Mathematik, `www.kfunigraz.ac.at/imawww/pages/humor/anekdoten_e.html`.

S. G. Krantz, *How to Teach Mathematics*, 2nd edition, American Mathematical Society, Providence, 1999.

J. E. Littlewood, *Littlewood's Miscellany*, edited by Béla Bollobás, Cambridge University Press, Cambridge, 1986.

D. MacHale, *The Book of Mathematical Jokes, Humour, Wit and Wisdom*, Boole Press, Dublin, 1993.

MacTutor History of Mathematics Archive, www-groups.dcs.st-and.ac.uk/~history.

Anecdotes about Mathematicians and Logicians, www.infiltec.com/j-logic.htm.

T. Pappas, *Mathematical Scandals*, Wide World Publishing, San Carlos, California, 1997.

J. A. Paulos, *Mathematics and Humor*, University of Chicago Press, Chicago, 1980.

G. Pólya, *The Pólya Picture Album: Encounters of a Mathematician*, edited by G. L. Alexanderson, Birkhäuser, Boston, 1987.

G.-C. Rota, *Indiscrete Thoughts*, edited by Fabrizio Palombi, Birkhäuser, Boston, 1997.

Science Jokes Collection, www.xs4all.nl/~jcdverha/scijokes/joketalk.html.

St. Andrews History of Mathematics Archive, www-groups.dcs.st-and.ac.uk/~history/Mathematics.

Top Ten List, math.ucsd.edu/~kowalski/kowalski/funny/mathjoke_pure.html.

Stanislaw Ulam, *Adventures of a Mathematician*, Scribner's, New York, 1976.

A. Weil, *The Apprenticeship of a Mathematician*, Birkhäuser, Boston, 1992.

N. Wiener, *Ex-prodigy*, Simon & Schuster, New York, 1953.

——, *I Am a Mathematician*, Doubleday, New York, 1956.

Index

Photographs appear in boldface.

L

M

N

O

P

R

S

T